Becoming a Firefighter

The Complete Guide to Your Badge!

Fire "Captain Bob"

Fifth Edition

CODE 3 PUBLISHING, PLEASANTON, CALIFORNIA

Becoming a Firefighter
The Complete Guide to Your Badge!

Also by the Author

Fire Up Your Communication Skills   *http://eatstress.com/fireupff.htm*

Eat Stress For Breakfast   *http://eatstress.com/stressfire.htm*

Published by
Code 3 Publishing
5565 Black Avenue
Pleasanton, CA 94566

http://eatstress.com

ISBN 0–9657620-0-9

Printed in the United States of America

Dedication

To those who possess the burning desire

to acquire a firefighter badge and become

one of the last of America's heroes.

Who is Fire "Captain Bob"?

Fire "Captain Bob" Smith has coached countless entry-level and promotional candidates to get their badge. Thousands of candidates have received their badges from this program. Captain Bob is a retired, 28-year veteran firefighter and lives in California.

Captain Bob is a recognized speaker/author on job interviews, author of the CD/DVD program *"Conquer the Job Interview"* and the books *"Becoming a Firefighter"*, *"Eat Stress For Breakfast"* and *"Fire Up Your Communication Skills"*, which have been translated in 22 countries including South Korea, Latin America and China. He has been a rater on more than 100 oral boards, a coach, publisher, frequent talk show guest, **(completing over 300 media interviews, including the Barbara Walters Show "The View"), featured in** *USA Today* **and the** *Wall Street Journal.*

He incorporates his own experience gained from three successful start-up businesses, a 45-year marriage (29 years that were good according to his wife), education, and thirty years of research. His wife has asked us to please . . . please buy some of his books so they will have more room in the garage.

Contents

The Problem is *Poor* Oral Board Skills!

Most candidates do poorly on their oral boards. The problem is most of them don't know how poorly they are doing. I've seen it too often after being on over 100 oral boards. It's the most misunderstood and least prepared for portion of the testing process.

Bottom line, most candidates don't do enough interviewing to get good at it. This is also true for any job interview. You've got around 20 minutes for a 25+ year career. How are you going to stun the oral board panel to convince them to give you the badge over the other candidates?

> *You've said the oral board is the most important step in the hiring process and the step that is least prepared for. Well I can support your statement. As I watched the recording of my score for LA City (reading upside down) I noticed 60's and 70's on the scores of those that interviewed before me. I would assume there were about 25 names on the sheet prior to mine. Again the methods you teach do work. I scored high enough to go into backgrounds and get hired. Thanks again for your help. Rich*

With all respect to the following comment, this is one of the most important clues why candidates have trouble in their oral boards:

> *"I recently had an interview, and I know my answers were great especially after hearing how another candidate answered them. He made the list, and I did not. Go figure! Jed."*

This is the problem! Most candidates think their answers are great when they aren't. If their answers were as great as they thought, they would make the list and get a badge. They listen to other candidates and firefighters who make them into clones. Have you noticed that once a person becomes a firefighter, they're instantly the experts on how to get hired?

The following is a true story: If we can get the village idiot hired, you can get a badge too! That's right! We had a candidate who was a volunteer taking his third test to be hired as a paid member. When this guy was out of town, the village didn't have an idiot.

He received our program and did not one, but two coaching sessions. We literally held him together with crazy clue. Guess what? He got hired. The next week, he made this posting on a bulletin board, "I will show you how to get a firefighter job." The village idiot had become the expert overnight on how to get hired. I couldn't believe what I was reading. He received thirty e-mails.

If you're passing the written and agility, which are usually pass/fail and you're not placing high enough on the oral, that's where the problem exists. What most candidates do if they don't place high enough on the oral is go back and try to pack on more credentials. "Oh, I have to finish my degree or get through that academy." They do little to nothing in gaining the skills for the oral board, which is usually 100% of the score. If you don't do anything to improve your oral board skills nothing is going to change, you will never ever see that badge. The oral board is for all the marbles. This is where the rubber meets the road.

Stop looking in the magnifying glass at others . . . and start looking in the mirror at your self. That's where the problem is.

Candidates who get this far in the process usually get discouraged and tell me they feel like they have hit a wall. They don't know what to do next. Some of their friends (with fewer credentials) have been hired. They're frustrated and embarrassed. If it can work for the village idiot, it will certainly work for you.

This is an e-mail received from a candidate. This is how fast things can change:

> *I ordered your Entry Level Program. I did this after noticing many of your students' successful testimonies on the bulletin boards. I have many certifications including Paramedic. The only hindrance that I found myself with was not passing the oral.*
>
> *Since ordering your program, I was nailing the interviews. Getting hired over the auxiliaries at their own departments and with a heck of a lot less experience. Your program and techniques helped me excel past the other candidates. I even had one city fire chief personally call me at home to set up a Chief's oral, (had to decline, due to the fact that I was at orientation for another dept.). To make a long story short, nothing counts until you have the badge, nothing. **For all of the candidates out there that don't believe this, try passing and ranking #1 on orals with a stuttering problem . . . I did.** Thanks Captain Bob — Dave*

Remember: "Nothing counts 'til you have the badge . . . Nothing!" Ask Dave

Has any what you've read made sense? Would you go on an African safari without a guide? Then why would you go to an oral interview without a guide? Would you cross a river without a guide to show you where the rocks are so you can make it across the river without being washed away? Haven't you been beat up enough yet? We would like to work with you to turn things around. It's been said that when the student is ready to learn, the teacher appears. Are you at this point now? We can help you wherever you are in the process. From the written test, physical agility, resume, oral board, background, psychological, polygraph to the promotional interview.

We can shorten the learning curve to the closest point between you and the badge. Like the Tin Man in the *Wizard of Oz*, we're not going to give you anything you don't already have. We're just going to show you where it is. There is a badge out there for you. You just haven't seen it yet. We will show you how to nail it!

It's a great feeling if you can be a part of the change in someone's life. Countless thousands hired with this program throughout the United States and Canada. This is our reward. Our son Captain Rob and I have a great passion in seeing candidates get a badge. This is serious work.

"Do what you have to do be more marketable so you can take more tests and have something more to offer a department, but remember that it all comes down to that 15 to 30 minute oral interview. I've seen some awesome candidates with resumes packed full of accomplishments that couldn't sell themselves in an interview to even make the top 50%." Steve Prziborowski, Battalion Chief, Santa Clara County Fire Department.

"Being a firefighter is a front row seat for the greatest show on earth!"

Here we go. Keep your hands and feet inside the ride at all times.

Where Are You Stuck?

"Getting the job of your dreams is like winning the lottery!" Jerry Price, Firefighter

Why "Nothing counts 'til you have the badge . . . Nothing!"

Certainly there are other things in life that are far more important than a badge; Family, health, friends and happiness.

The reason I use this slogan is because I often hear candidates who don't have a badge use these excuses. I have education, experience, training, been a volunteer, bunked, and have a burning desire to get this job. Got degrees, certificates, have every merit badge everything you could think of, yada, yada, yada. None of this is going to matter if you don't get a badge. You will still be the bridesmaid.

Take this simple test to check how you're doing getting a badge:

l   Are you subscribed to a service that notifies you when a department is testing? Check out firecareers.com and Firehouse.com

l   Are you taking every test you can?

l   Are you passing the written?

l   Are you passing the physical ability?

l   Are you preparing for your oral interviews (100% of the score)?

l   Are you passing the oral?

l   Are you getting conditional job offers?

l   Are you passing the psychological interview?

l   Are you passing the medical?

l   Are you passing the background?

If you can't answer yes to all these questions, you will never see a badge! Wherever you answer no, you cannot go onto the next step to gain a badge. Where are you stuck? Check out the appropriate sections of this book where you are stuck to gain the information to get on with getting that badge.

I get calls from candidates all the time telling me about all their wonderful credentials. They have been testing 3, 5 or 7 years, have been number 30 on this list, 22 on that list, volunteer firefighter, AA in Fire Technology, some medics and every certificate and merit badge you could imagine. I have to stop them before they get into warp speed with all their stuff. I do this with one simple statement: Do you have a badge? They go off again with more of their great stuff. I bring them back with "But, you do not have a badge? You are the bridesmaid. Never the bride."

Many write e-mails like this: I know I have what it takes to make it as a full-timer.

My reply: Yeah, they all say that. You know that. But you have to convince the oral board panel that you really do. That's where we come in.

You would not be calling if there wasn't a problem? Right? Finally they answer, right. If you cannot be humble and ask for help, how can you convince them on the oral board you can be humble enough after you get hired? Once we are both on the same page, we can start working on where you are stuck and find a solution that will get you that elusive badge.

I receive e-mails like this one:

> *I am discouraged. I was in a recent psych/background process with a large department and a friend of mine got his conditional offer today and I did not. I am a new medic with six months experience, I have an AA, 14 units shy of a bachelors in management, I was in the Marine Corps, I am 34 years old and I felt like my interview with the psychologist went well and I am unaware of any major background problems. What gives? My friend is 22, no college and brand new medic himself. I just don't understand . . . — Steve*

A candidate like this called us last year. Skeptical, he went ahead and got our program and did the private coaching. He called yesterday that he had been offered the job of his dreams. He told me when he took the oral for this department there was a battalion chief on his panel that kept his head down writing most of the time.

When he gave his nugget signature story answer to one question, the battalion chief looked up, put his pen down and focused on his personalized answer (a real good sign). He said he knew then he had the job. Dan went to his psych armed with our special report on how to prepare. He said it wasn't difficult avoiding the land mines during his psych because he knew where they were (this was probably the problem area with the discouraged candidate above). Dan got his badge. He had rushed out after the ultimate call to take flowers to his bride-to-be. He started the academy two weeks later.

Firefighter or Fire/Medic?

Should you become a paramedic to get a firefighter job?

You understand that there are up to 800 candidates for each firefighter job, that you could be running out of hope and the love of your life is not going to wait any longer unless she has a ring and a date. Been the frequent-flyer mileage king flying all over the country and unsuccessfully taking tests and your biggest income last year came from your credit cards. You have lost friends. Don't know how you're going to live without the job of your dreams. You can afford to take the time and loss of income to make it happen. Have a supporting partner. Know you would have to spend about a year getting certified and it will be the toughest thing you have ever done.

Can a relative pay for your education? Do you have GI Bill education money? Can you get a student loan? Know that 80% of the job offerings now are for fire medics and up to 75% of our calls are EMS related anyway. Know that even if you become a medic, you still may never get this job. Have been riding ambulance anyway and this will pave the way into many medic schools. Aren't going to take the chance of some college medic programs that only take 35 people a year and receive over 200 applications. Will step up and pay the $9,500 plus dollars (student loans available) to be in and out of a program in about a year.

You're the energizer bunny who will keep going and going and going when others would stop. Know that if you are a medic taking a regular firefighter entrance test you will probably get a better shot. You won't be happy until you can puff your chest out with a badge and have people wave at you in the jump seats, carrying on a family tradition. Want that shift work with great benefits that go way into retirement. A career position with chances of advancement.

You will have the opportunity to use the education and experience you have acquired. To work for a department that offers you everything a firefighter hopes for. Calls that cover anything from air, land and sea. A place where you can't wait to get back from your days off. You will be able to go from one call to another to another on a busy rig. Believe me there is nothing like it.

I know you will hear that if you really don't want to be a medic don't just do it to get the job. That all you really need is your EMT to get hired. But, if you answered yes to the majority of the above there is no doubt where you will be the happiest. If I were in your position I would beg, borrow, and run with my afterburners on to get to medic school! Because unlike a regular entry level test where there are up to 800 candidates for each job, there are only 20 candidates for every fire medic job. It is by far your fastest way to the badge.

"Even being a paramedic doesn't guarantee you a job. You still have to pass all phases of the hiring process, and most importantly, pass the oral interview. You can have all the certificates and education in the world, but if you can't sell yourself or effectively communicate your qualifications in the oral board, all your pieces of paper are useless." Steve Prziborowski, Battalion Chief, Santa Clara County Fire Department.

Lucky Joe?

With all this said about becoming a medic, we help candidates get hired all the time with little or no experience. Introducing Joe:

Joe got one of our flyers when he picked up an application for the San Francisco Fire Department test. He had seen the ad for the test in the newspaper. Although he was not a kid, he thought he would give it a shot. No background, formal education or experience related to the fire service.

It didn't take long after Joe contacted us to figure out that he was not the brightest light on the tree.

We suggested a book he could get to help him prepare for the written. Every time he called me he also called my firefighter son Rob. On one call he left a message asking, "If I buy this book, does it have the answers for the San Francisco test in it?" There were many, many calls. We answered all.

Of the 4100 that took the San Francisco written, guess who was one of the 1700 who passed? Joe passed the SFFD written.

The oral was next. My son Rob dreaded doing the coaching session. It normally takes about an hour for a coaching session. It was a looooong session.

Guess who was one of the 609 who passed the San Francisco Oral? Yep, Joe.

Next was the physical agility. Joe was in great shape. He wanted to know how to prepare. I told him to go to the Physical Agility section of the FREE "101 Inside Secrets to Get a Badge" on our web site. Joe left a message the next day, Captain Bob I can't find that book you told me about on preparing for the physical. Another call to Joe that it's FREE on the web site.

Joe doesn't know about you. He doesn't know about your degrees, certificates, education, and experience.

Even though Joe is not the sharpest knife in the drawer he did something you probably aren't doing. He said, "Captain Bob just tell me what I need to do." And, he listened! He listened to what he needed to do. Joe followed the simple formula: Get the program. Fill out the work sheet. Work it. Use a voice recorder to practice. Get coaching. Badge!

Guess who graduated from the 108th fire academy in San Francisco? That's right Joe. He walked right in off the street and took away your badge.

I told this to a medic testing for 4 years in the Denver area and he said, "Captain Bob that's not funny." I know it isn't. But Joe is now wearing the badge that many aspire to wear.

Is it just a myth?

I wanted to know why this one candidate was in the hiring process. He had no academy, no college degree, no EMT, and no experience. I thought to myself, how could this guy be in the same group as me?

Reply: That's the myth. Candidates believe they have to accumulate a bunch of credentials to be hired. This alone will get them the job. The truth is it's not what you have or don't have but how you present those credentials. The rubber meets the road in the oral board.

Don't get me wrong, credentials are great, but we've had numerous candidates, like the one you described above with few or no fire related credentials that get hired. They realized they couldn't compete with candidates with overwhelming credentials. So, they improved their chances by concentrating on the most important part of the process. THE ORAL BOARD! They converted their personal life experiences into proven oral board skills needed to get that badge! Like you, this baffles the other candidates. They can't or won't believe it can be done. Right now we have several of these candidates with seemingly no credentials in the hiring process in major cities across the United States and Canada.

Question from a candidate: Are you trying to tell me that a city would hire a candidate with no fire education or experience? No way!

Response from a candidate who did it:

There is a real simple answer to your question . . . most, if not all big city departments require almost none, if any, experience or certifications. Says so right on the job announcement. Unless they are required by the department to test, certs and experience mean squat.

The people who get the jobs are those who can show during the oral board process that they possess the personality, willingness to learn, ability to adapt, and how all of their past life and work experiences have made them well suited for the career. These are the people who blow right by the other wannabes.

You have to be pretty blind and ignorant not to understand why the bigger departments run their own academies . . . so they can teach the recruits "their way". None of these departments care one bit how it was done at whatever academy you have been to before.

Captain Bob shows you how to handle the situational questions and how to demonstrate to the panel members that you can and want to do the job.

Even with no fire experience and just a lowly EMT cert, I had 4 conditional job offers over a two-week period. My secret? . . . Captain Bob!!!!!!

Education

Which Path?

I'm a college student finishing up my AA in Fire Technology. I have talked to many firefighters most of which informed me of the following:

1.An AA in Fire Technology is sufficient for being a firefighter
2.An AA in Fire Tech will be enough with EMT and CPAT to get you into an academy
3.Is it better to get a bachelor's in something other than fire technology or learn a specific trade?

The question I have is number three. Is it true? What other major or trade would make a wise decision?

Ask yourself who is getting the badges? The vast majority of candidates we see get hired do not have advanced degrees. They're more in the line of EMT, FF1 academy, working on or have an AA or AS degree or medics. Some have no fire education or experience. Their biggest asset was they leaned how to take an interview.

First leave no doubt that I believe in education. If you want to get a Public Administration, Engineering or any other degree as a career track, great. Don't think it will be the key to get into the fire service to ride big red. We hire candidates not credentials on resumes.

But where are you going to get the most bang for your buck? We have enough chiefs. We need more Indians.

The following is from: Michael J. Ward, MGA, MIFireE
Assistant Professor, Fire Science Program Head Northern Virginia Community College
Annandale, VA http://home.gwu.edu/~mikeward

In my preferred world, a high school graduate will attend college and obtain an undergraduate bachelor's degree PRIOR to getting a "real" job. This illustrates the values of going to college and getting to experiment and become an adult in a semi-protective environment.

Lets cut through the testosterone and turf wars and consider the question of which is the best way to get a badge. First, I will agree when considering a major in college, fire science provides a poor return on investment if the goal is a career as a paid firefighter.

Firefighting is one of the few middle-class jobs not requiring college education as a pre-employment requirement. I think that distinction will evaporate in the next generation. As Captain Bob repeatedly points out, most fire departments do not provide preferential considerations for someone with a two-or-four year degree. If you are going to college to prepare for a career in fire-rescue, your best investment is to obtain paramedic certification.

Everyone has an opinion, there are exceptions and more than one road to a badge, and there are no guarantees in life which ever path you take. Education will never hurt you.

If you really want to get a firefighter job consider these points:

Is there a requirement for an advanced degree to get a firefighter job?

Answer: It's rare to see departments require an AA or an advanced degree to apply.

Where are 80% of the job offerings?

Answer: Fire/medics

There are up to 800 candidates chasing each firefighter job. How many are chasing a fire/medic job?

Answer: 12-20. Which odds do you like better?

What's the time-line? If you're just starting college and want to get your BA, it could take you 4 maybe 5 or more years depending on when you can line up and complete all your classes and requirements. Then, if you wanted to go further timing it to get into an academy or paramedic school and get some street time another 2+ years? So around 7 years give or take to get in position to go after the badge. Are you going to need student loans? Do you have a special person in your life that is going to wait while you pursue your career? How long can you tread water?

The path to become a medic is about 2 years with gaining some savvy street time. If you can get in an academy in that time period it will be convincing evidence that you have the hands-on experience that a department can take a risk on you.

Can you continue your education once you're hired? Will departments give you an education incentive?

Answer: Yes to both.

Yes, having a degree will help with promotions but how long will it be before you will qualify to take a promotional exam?

Answer: Engineer depending on the agency 3 plus years. An officer? Five or more years. So if you get on you could obtain the necessary education before your first promotional test to be in position. And, the department will pay for you to go to college.

Remember, You're Just a Rookie

I've coached candidates who have had B.S./BA degrees in Public Administration areas. They have been misguided by counselors that said this would be an asset to get into the fire service. What ends up happening is these candidates show up at an entry-level oral board boasting and trying to hammer the board with their degrees.

John came in for a coaching session after not being able to pass any oral boards. He was one of those candidates who I think was misguided into a Public Administration Degree. During his coaching, he kept trying to come back to his degree. I finally told him, "Screw you! You want to come into my oral board and try to hammer me with a degree you may never use?" You're applying for a snott-nose rookie position as a firefighter!" John dropped his head and said, "Maybe that's why I can't get through any orals."

John ended up going to paramedic school (which he should have already done instead of the B/A degree). Although he mentioned the B/A degree in his oral board answer "What have you done to prepare for this position" he focused on his personal life and paramedic experience. He got his badge!

> *I am graduating in March with my Bachelor's in Fire Services Administration from Eastern Oregon University with nothing but Intern Firefighter/EMT experience. I didn't learn anything about being an entry level firefighter during the process (I learned all about that when I earned my Associates in Fire Protection Technology), but what I did gain was a vast amount of knowledge of the administrative side of the fire service, which I was previously shielded from as an Intern. I was given great opportunities to work alongside seasoned firefighters, chiefs, captains, and lieutenants on projects, assignments, and discussion boards. The advice I gained and the relationships I built with the seasoned firefighters were very beneficial. Hawke*

From Tom Dominguez: Don't make the assumption that having a BA/BS in any discipline entitles you to a fire job. Even if you have a BA/BS in a fire related subject, you are not entitled to an entry-level fire job. If you think a BA/BS degree can be exchanged for a fire job, you do not understand how to take an oral interview.

From: Captain Rob http://www.myfireinterview.com e-mail: mailto:captrob@sonic.net

Learning about fire administration and being around the upper brass of the fire service is nice. But I will say your time is better spent becoming a good new firefighter. We have been saying for years, becoming a strong EMT and getting you medic is far more important than understanding how the budget is put together. We aren't looking to hire a chief.

Those skills you learned working in admin will not help you get through an interview, the academy, or your probation period. In fact it could hurt you. Bringing up advanced management experience and degrees, and how you have been there and done that can doom an interview. And it can also tank a probationary period if you don't know how to leave that stuff in your locker. But being about to show command of a medical scene, a working knowledge of being a new guy is what will get you a chance to show your stuff and get through the process. That is not to say a degree is a bad thing, but there are definitely things you can do to make yourself more "hirable" before you move on to advanced degrees.

A person with a few years as an EMT then medic and a little volunteer time will have a much easier time getting a job offer than someone with their credentials, and a B/A and no experience.

If you were to get your EMT, F/F1, then medic, then start testing, you could be taking classes for a higher degree while you test. If it takes a year or two to get hired you would then have just a few classes to finish up after you are off of probation. That is if it is still important to you at the time. We have had at least 50 people complete their B/A after they were hired with my department.

From Jeff:

> *With all due respect to all that was said, speaking as a volunteer firefighter who has a 4 year degree, I would say that getting your BA or BS for a firefighter job is not a good way to go. I got my BS, and $100,000 later, I'm hoping to work in a job that requires only a technical certification that costs $250 and having the BS without the tech cert makes me pretty much unhirable.*

> *Get your paramedic. Get your FF1 and FF2. Get your hazmat tech. those are what is going to make you valuable to a company. A 4-year degree is worthwhile, but only if you use it. Most departments aren't requiring them. Some departments will even pay for you to attend college courses. Yes, it helps if you want to become an officer or a chief officer. But you're going for entry level. You're going to have to pass the physical, pass the psych test, pass the written, and pass the oral board.*

Focus on your goal and don't let anything get in your way until you get it.

Is Age a Factor in Hiring?

I've coached lots of candidates over the age of 30. It's not uncommon to see candidates in their late thirties or early forties hired, especially paramedics. I encourage candidates to focus on their personal life experience when answering questions in the oral board. No one else can tell your stories of where you have been. Candidates over 30 years old have life experiences younger candidate can't match.

When candidates answer questions laced with a story that demonstrates they have lived the experience, they separate themselves from the other younger clone candidates.

I work shoulder to shoulder with candidates through the process until they get the badge. Here are testimonies from a couple over 40 year-old energizer bunny candidates that kept going and going and going when others would quit:

Subject: I made it !!!!!!!!!!!!!!!!

Hello my friend,

Well after 19 years of trying and testing and reading and wanting, I made it. At the ripe old (I mean young) age of 41 and a grandfather of 2, I was accepted on to the fire department.

I cannot begin to tell you how I feel after so many years of trying to reach my goal of being part of the CSFD. No other department would do (except as a stepping stone) it had to be here in my hometown and I made it.

I wanted to send a special thanks to you for helping me reach my goal, not only with your Entry Level program and books, but also for the encouragement along the way.

You have helped this old man to be a very happy one and I appreciate that in you.

Take care, — Steve

Drum roll please: You're not going to believe this. Don't miss the last sentence!!!!!!!

Indescribable Thanks Captain Bob:

I have arrived at the moment, that before now was only realized in my dreams, that was the day I received "the call" informing me that I had been chosen as one of the fourteen candidates to start the academy for Las Vegas Fire and Rescue!

I am very fit. I work out 5 days a week and have for years. I have my EMT-Intermediate Certification and a burning desire to become a firefighter since I was 10 years old. Some times life seems to get in the way of our dreams and plans, but I have learned only if you let it! Three years ago I decided to go for my dream career, I had done everything I knew I could do to physically be ready for this demanding job, but I was oblivious to the testing and hiring process. That is when I came across your web site. Captain Bob your materials were invaluable in helping me obtain my badge. I followed your advice to the letter throughout the process from testing to oral interviews and the psych. Your insight is right on the mark!

Rob helped me with private coaching, and low and behold I am poised to start my dream career! "Thank you" is not enough to express my heart-felt gratitude for your help in getting me to this place in life. I look forward to meeting you in person some day to say "thanks". Ted R. Las Vegas Nevada

Ted was number 3 on the list, an EMT and FORTY-SEVEN YEARS OLD! He also scored higher than his son on the same list and got the job. Bravo!

Younger Candidates

Younger candidates have credentials too!

Question: Just recently, I began testing for fire departments in California (predominantly southern California). I am a fire academy graduate and I have two years of experience on a basic life-support ambulance.

I am also currently testing for auxiliary and volunteer positions for added experience and my firefighter 1 certification. In January, I'm planning on going to paramedic school. My question is this... I'm only 20-years old- I won't be 21 until August. Does being this young hurt me whether or not I shine on my written and oral tests?

I don't see too many rookies this young with departments and I don't hear about people being hired this young. LCM.

Reply: As long as you can present your package at the oral board, age should not be an issue. The problem is many younger candidates don't think they have the life experience needed. Never tell the board your age.

LMC, I would continue your pursuit for a volunteer position and definitely get into medic school ASAP.

Many students at Shasta Fire College didn't feel they had any experience that would apply to the position. That was until I asked several candidates to tell me about their first and succeeding jobs in life; no matter how menial it seemed. Many had paper routes, mowed lawns or worked at Burger King. O.K., what did you learn? Once the answers started flowing, we heard how they learned to work hard, have responsibility, provide customer service and how to work as a team. Did you participate in sports in school? Isn't that working as a team? Do any of these areas apply to the fire service? You bet! So anytime you can relate your personal life experience in answering an oral board question, you are telling the oral board that you not only know the answer the question, you have already lived it!

When the board asks what you have done to prepare for the position, don't forget to rewind the videotape of your life and create an early trail of how you learned how to work hard, have responsibility, and work as a team.

The biggest part of getting a high enough oral board score that will get you the badge is convincing the oral board you can do the job before you get it. Stories are convincing evidence that you are the match for the badge!

I had several Fire Explorers who were too young to test. They handed out flyers at written tests in exchange for products and coaching. At an Oakland fire badge ceremony, one of the fire scouts got a badge on the first test he was old enough to take. You have never seen a happier rookie firefighter.

Applications

What's the first impression the job panel has of you? Your physical appearance? Yes. What else? Your choice of words, eye contact, and your handshake? All-important. You probably missed the most important point!

Your application and resume precede you before you walk in the room! I can't tell you how many times we've seen applications with misspelled words, chronological order wrong, and we haven't even seen the candidate yet!

If you have the opportunity to get an application ahead of time, make a photocopy. Some agencies let you download the application off their website. Plug in the all information on the copy. Have a qualified person review and correct it (your mother, girlfriend or wife).

When my son was trying to get on the fire department, he had his mom do that. She is a good speller and a good typist. Put everything down. Then you've got something that you can transfer to the real application, and that becomes boilerplate. You can use it any time you have a new application. Many applications now are computer generated. It is difficult to type your information into the limited space.

Some applications are a simple Scantron that you can fill in on the spot. Some agencies have you fill out longer applications on the spot. Make a photocopy of the completed application before you send it in. You never know when you're going to that job interview. I talk to people who have put in applications and six to eight months later, they don't have a copy and don't remember what they've put down.

Bad Stuff on Applications

If you do not include information that is asked for on an application and it is found out later, you are out of the process! Almost everyone at sometime has problems. It's how you put them on the application, background forms, and present them in an oral that makes the difference. A reasonable explanation is what's important.

Many candidates strain their relationships, marriages and finances and do various jobs trying to get the badge. This is understandable with the right explanation. The oral board seldom knows this information (this is usually covered in background), unless it is an area that is listed on the application, i.e. driving record, arrests, etc.

I served 5 days in Santa Rita Prison for drag racing at age 18. Yes, I put it on my application. Because if you don't and they find out, you're gone. In my oral board, I was asked about this. I told the panel, "Since that incident, I have been in the army, married, have children, and have been on my job for 9 years. I was a stupid kid. The situation hasn't occurred again. It's hard to believe this really had happened. One of the captains asked, "Mr. Smith are you trying to get go around this problem and ignore it?" Here's the Nugget answer: I said, "No. If I was trying to do that I would have never put it done on the application." He was done with that question.

When I got my results for that test, the number placement wasn't on the notice. When I called, personnel told me, "Well, Mr. Smith, you're number one. Not only are you number one, you're five full points ahead of number two!" It was having a reasonable explanation prepared in advance that becomes your "Nugget" answers that makes the difference.

That question and the "Nugget" answer helped me, not hurt me. It catapulted me past the other candidates at light speed, and did indeed help me get my badge!

Pulling Out All the Stops

Candidates were asked to submit a photo with their applications to go forward in the hiring process. One candidate called to find out what the requirements were for the photo. They said just a head shot, because one candidate had already submitted a photo of himself in uniform, standing on an engine, with a Dalmatian dog on a lease, and holding a little girl in his other hand. Talk about pulling out all the stops.

An oral board rater told me they had interviewed 175 candidates. When it came time to make the selections, the two applications that included photos made the final cut. He said the photos drew them back to the candidate.

Preliminary History Questionnaire (PHQ)

Many departments are including a hand out or online Preliminary History Questionnaire (PHQ) as part of the application process. The PHQ is designed to eliminate candidates early on in the hiring process before they spend lots of time and money on those who they would tank later.

So you take an on-line PHQ and when you try to continue it flashes PERMANENT DISQUALIFICATION! Or, there Could Be a Delay in Your Background. What happened?

Well, you probably answered correctly or incorrectly yes to a question that had a threshold that was an automatic disqualification. You still might not be out of the woods on some of your answers. They could come back to haunt you later in the process. Make sure you go back and double check your answers before you click continue. You would be surprised the percentages of candidates are being disqualified from this process.

There is no question that candidates should lead their lives as if their future consideration of being hired as a firefighter depended upon it. A department may include in the background information packet that a polygraph may be used to verify the information you submit.

Threshold

So where's the threshold? Did you answer yes to a question on drug use? How many times? Answer yes to a question on how many times you drank more than the legal limit and driven? Driving or criminal record? Bad credit?

So what do you do?

Learn from you're experience. Before you submit future on-line or hand out printed forms double check your answers for errors and think twice before you step on any landmines that could take you out of the process.

What others are doing?

I finally figured out that if I listed those things that happened in my life a long time ago I could be eliminated from the process. I asked myself if I didn't tell them, how would they find out? Steve

This PHQ nailed me the first time I answered yes that I tried cocaine 3 times 20 years ago. I didn't have a chance to give an explanation as I have had sitting down with a background investigator. Dennis

Last test I filled out the PHQ hand out. I answered yes to the question have you ever had sex with a minor. I was rejected, even though I wrote a half page explanation that I was 18 in high school and my girlfriend was 17. Thinking about it now how would that ever find this out? I'm now 27, no longer live in that state, and have no idea what happened to my girl friend. I can't imagine how they would find her and if they did can't imagine she would tell them out of the blue we had sex when she was at age 17. Andy

Why did I tell them I smoked pot 20 times 10 years ago in England? We're they going to fly over there and find my old friends? Dan

My question for you is in regards to the "Bad Stuff on Applications." I have your book and other materials in which it states, "Do not open a can of worms." However, in this email it states that "If you do not include information that is asked on an application and it is found out later, you are out of the process!" Which is the correct way to go?

For instance, I got into a fight as a youth (17 years old) protecting my brother and had to go to court where I was found not guilty and had to perform community service. Since it was almost 10 years ago and I was a minor, is this something I should disclose on the application? That is the only instance where I have any possible mark on my record, including traffic citations (or lack of). Thank you in advance for your time, Rob

Although juvenile records for minors are generally sealed, the question asked if you ever; yes or no? I would put it down. If you put no and it is revealed later, you could be out of the process. This is something that should have no affect on you being hired by a department. What is most important is that you have a reasonable explanation of what happened and the court found you not guilty.

Which is the correct way to go?

First of all what I was referring to is if there is no trail, in other words who can they talk to who would know this information? If there was an arrest, a court appearance and community time there is probably a record of that somewhere. Then why open a can of worms by creating one. Volunteering information that was not requested eliminates way too many candidates.

You're a free agent. Make sure you prepare for the hiring process in a way that will best put you in a position for a badge. Steve

I can't tell you what to do. I present the facts and you get to decide.

Unrelated Degrees

How to present your unrelated degrees?

I have three degrees in law enforcement. Every time I go to an oral board, they seem to target and spend too much time discussing my law enforcement education. What can I do?

This is a problem that many candidates face with degrees that are not fire related. The board will see this and follow the rabbit down the hole; spending time having you defend your position. How do you even begin to relate a degree in "Romance Languages, Art, Philosophy, Economics, etc." to the fire service?

Don't create the trail. Start listing just your degree on your application, resume and including it in your oral board answer as what you have done to prepare for the position. However, omit the field it's related to, i.e., Bachelor Science Degree (leave off Law Enforcement). Don't bring it up and they probably won't. If they do, tell them you learned how to learn. A study from University California at Berkeley showed that two years after graduation, 60% percent of graduates were in fields totally unrelated to their degree.

An additional question:

I have been facing the same problem about my degree. One chief told me to bring up my degree, how I have used it and how it brought me to a career in the fire service. He told me to not let the oral board bring up the degree because once that has happened they have already formed an opinion about it. He told me to include it in my history and tackle the issue before it becomes an issue. For example, you have a degree in Law Enforcement. This has allowed you to work with firefighters from another angle and you see how the fire service is and that is more the career path for you. Hope that is helpful.

Reply: Have you tried this approach? It might work for you, but it is our experience that once you open the can of worms, it's impossible to close. The explanation of being in law enforcement has enabled you to be around firefighters is pretty weak. Many would agree that firefighters and police officers are cut from a different cloth.

I still believe not creating a trail they can follow is the best solution. Our candidates who have left off the field from the degree have had few if any problems.

Here's another possibility:

I have 2 Bachelor degrees, one in Finance and one in Accounting. I am currently working as a full-time firefighter with a good-size department. You can overcome this "focus" on your degree by accentuating the positive . . . that you were able to dedicate 4–5 years of your life to obtaining the degree. You sacrificed, set a goal, and achieved it. That is what is important, not what the degree is in. Try that at your next interview, it may help. Being a firefighter was not a life-long dream of mine, I changed my career path. How can anyone fault you for wanting to educate yourself, or choose a different path, one that makes you happy? Good luck.

Resumes

Most resumes are poorly done. The business resume format is not the best for firefighter candidates, because with the high volume of candidates, the raters only have a few moments to look at your resume before you walk into the room.

I'm a one-page resume guy for entry level without a cover letter, not in a binder or folder. Do not give us a book. We will not read it. The board does not have enough time. And do not come into my interview, any interview, thinking you are going to hand out your resume and we're going to read it. That is not going to happen. This upsets the normal flow of the interview. We're going to read your application and resume before you come in the room. If you submit a resume, get it to personnel to be placed in your file before the interview. Don't fax It. Make the appropriate copies and hand deliver or FedEx them.

A candidate faxed me his resume for review. The cover letter for the position he was applying for stated, "Attached is a **"brief"** description of my qualifications." I laughed out loud because he had sent me a book. The printer ran out of paper. Save a tree, the raters will not read these volumes. Don't send me on a treasure hunt to find your great stuff. Hit me with your major qualifications starting with your experience on one page. Write it believing the raters won't go past the first page. You can put any supporting details, documents, certificates and letters of recommendation following the first page. Keep it simple.

Many people start their resume with their education. For me, I like to see professional experience jump right off the page. Hit me with experience, bam! Fire fighting, bam! Some kind of training, apparatus operator training, fire school, whatever it is. Hit me with that experience. And that doesn't necessarily have to be in chronological order or fire service experience. On so many of the resumes I see, I find the important stuff way down at the bottom of the first page. Because that's how it falls in chronological order. It starts with some education up here, some college, whatever, blah blah, experience, now we're down at the bottom of the page where I might not see it.

I was reviewing a candidate's resume and in chronological order his paramedic certification was at the bottom of the page. I asked him, "What are the most important items on your resume? He said, my Firefighter 1 and Paramedic Certification." They were at the bottom of the page where they might be missed. We put those items on top so those are the first things that hit you. We put the dates on the right side of the page where it can be referenced. Once you put the dates on the right-hand side of the page, you list your experience in order of importance, not just in chronological order. This makes a big difference.

My suggestion for a firefighter resume format: name, address, phone number & e-mail address, professional experience, education, volunteer and community service. That's all you need. Nothing more. Nothing less. Keep it simple.

Make a photocopy because you never know when you're going to that job interview. I talk to people who have put in applications and resumes, and six to eight months later they don't have a copy and don't remember what they've put down.

The next two pages contain the before and after of an actual resume we received (the name has been changed) from a candidate. After seeing the revision, the candidate wrote, "Captain Bob, wow, what a dramatic difference. Much easier to read and pleasing to the eye."

I can review your one page resume to make you look like a pro too! First though put your resume in the format on the sample page. Save it in a Word.doc, not Vista and e-mail your resume to mailto:robert..smith19@comcast.net , I will suggest changes send it back to you. Cost $20.00.

Well I'm sorry to say that at 37 years old I thought I knew how to put together a resume. I sent an e-mail to Captain Bob with a simple question. He asked me to fax over the resume for his review. OUCH, man did that hurt!! It came back with all kinds of "who cares" and scribble marks all over it.

Well you know what? I have never had a resume look so good in my life and with two pages and years of classes, I thought there was no way it could be one page. Boy do I still have a lot to learn! I advise you to give Captain Bob a chance to look at your resume too. You might just be surprised by what you see. Thanks, — Allan

I never realized until now how hideous my resume really was. I don't need my resume very often, but when I do it's nice to know it's mint! I can be confident in that now. Thanks again for your services. — Kyle

Carl Mcfly (Original)
1284 Main St.
Kensington, Ca 94588 Phone: 510-286-5890

OBJECTIVE
To gain employment with a Fire Department

EDUCATION
Bachelor's Degree, State University, CA 1996

EXPERIENCE

Firefighter   Fire Department, CA   2-00   Present

Duties include but are not limited to fire suppression in structural as well as wildland environments and emergency medical services under highly stressful emergency conditions. Also, fire prevention, public education, vehicle and station maintenance under the supervision of a Captain, always focusing on providing quality customer service.

May 5, 2000 EMT Defib and Combitube certified

August 28, 2000 Firefighter I

August 28, 2001Firefighter II

July 17, 2002 Member of Federal Red Card System for campaign and complex fires as Fire line EMT and FF2

July 15, 2001Class B Driver's License

December 15, 2002 State certified Driver/Operator

January 22, 2003 Red Cross certified CPR Instructor

February, 2003 Made Engineer Promotional list

Owner/Operator   CPR Training,   3-03   Present

Adult, Child, & Infant CPR training for the community as well as for the professional rescuer.

Auxiliary Firefighter   Department, CA   3-99   2-00

Perform in a probationary capacity under emergency situations, fire suppression, emergency medical services, also fire prevention, public education, vehicle and station maintenance under the supervision of a Captain, and always focusing on providing quality customer service.

EMT   Ambulance Service, CA   3-99   2-00

Perform under emergency situations, emergency medical services under the direction of Redondo Beach and L.A. County Fire Department Paramedics. Vehicle and station maintenance and Code-3 driving.

Owner/Operator   Pool Company, CA   5-97   2-00

Service and repair of residential and commercial pools and spas according to County Health Department specifications and regulations.

ACTIVITIES
Member of University Track and Field Team.
Member of University X-Country Team.
Volunteer for Hubbs Institute White Sea Bass Population Restoration Project.
Volunteer for Red Cross on various projects.

Carl Mcfly (Revised)
1284 Main St.
Kensington, Ca 94588
Phone: 510-286-5890 e-mail: Iwantafirejob@aol.com

OBJECTIVE: To achieve a level within the fire service.

PROFESSIONAL EXPERIENCE:

Firefighter Fire Department, CA   2-00 Present
Duties include but are not limited to fire suppression in structural as well as wild land
environments and emergency medical services under highly stressful emergency
conditions. Also, fire prevention, public education, vehicle and station maintenance
under the supervision of a Captain, always focusing on providing quality customer
service.
Engineer (Acting)   2-03 Present

Firefighter Fire Department (Auxiliary)   3-99   2-00
Perform in a probationary capacity under emergency situations, fire suppression,
emergency medical services, also fire prevention, public education, vehicle and station
maintenance.

EMT Ambulance Service, CA   3-99   2-00
Perform under emergency situations; emergency medical services under the direction
of Redondo Beach and L.A. County Fire Department Paramedics. Vehicle and station
maintenance and Code-3 driving.

INSTRUCTOR Emergency Response CPR Training   3-03 Present
Adult, Child, & Infant CPR training for the community as well as for the professional
rescuer.

Owner/Operator Pool Company, CA   5-97   2-00
Service and repair of residential and commercial pools and spas according to County
Health Department specifications.

EDUCATION

Bachelor's Degree	EMT Defib and Combitube certified
Firefighter I	Red Cross certified CPR Instructor
Firefighter II	Federal Red Card System Member
Driver/Operator State certified	Class B Driver's License

ACTIVITIES
Member of State University Track and Field Team.
Member of State University X-Country Team.
Volunteer for Hubbs Institute White Sea Bass Population Restoration Project.
Volunteer for Red Cross on various projects.

Recommendation Letters

Question: Will it help to present letters of recommendations from prominent people, such as a former police chief, academy instructors, my priest or other firemen in the area? Will it carry any weight at all?

On most oral boards, the raters are from other departments. It is my experience that although the raters might thumb through and glance at any attached documents they seldom if ever read them. And come on, if you're going to attach a letter of recommendation, they're easy to get and it's not going to say anything bad but only glowing words about you.

Save a tree, the raters will not read these volumes. Don't send us on a treasure hunt to find your great stuff.

I'm a one-page resume guy for entry level without a cover letter, not in a binder or folder. Do not give us a book. Keep it simple.

Question: If you are going to attach any letters of recommendation following the first page how many is appropriate. I am thinking maybe two. I realize that there is a very good chance they won't read it but wouldn't it still look just a little better to have a couple of them?

Reply: How about none! Again, I'm a one-page stand-alone resume guy. Don't give me a book. And, if you have already listed your education, experience and certifications on your application and resume, why do you need to attach them; unless they were requested or you're going to a chief's interview where there is more time with each candidate.

Many entry level and promotional candidates have told me they were complemented on having just one, as in one page only, resume.

As you know everyone becomes an expert on these issues. They will fill you head with all these crazy ideas and stuff. And because "They said (I've been trying to find "they" for 30 years) you needed to have all that extra stuff or the other candidates are going to get ahead of you." So, how is all this extra stuff going to really help you?

Candidates have been told by firefighter friends to go ahead a hand out a resume with recommendations letters to their oral board panel even though the instructions stated they were not accepting them. Because, it would make him stand out. It did. But not in the way he was expecting. Keep it Simple.

Written Test

Having Trouble with the Written?

"Luck is given to the prepared", Thomas Jefferson

If you are having trouble passing the written test, try to find out what portion you are having problems with. Is it the math, mechanical, aptitude, or word comprehension? You don't need to take a college course to get ready.

Still the top program for getting ready for the written test is the Encyclopedia of Firefighter Examinations available on our web site. You can take a sample test with answers on our site: http://www.eatstress.com/donmcnea.htm

Of all the books out there for the mechanical aptitude, the ARCO Mech. Aptitude & Spatial Relations book will give you more than you will ever need. Perfect Firefighter has it at http://www.firecareers.com/

Try this approach. Instead of taking the sample tests in the written test books cold, go to the answers in the back and go through the first time with the answers. Then you'll know what they are looking for in the answers. It will cut your learning time. There are only so many ways they can ask a question on the same topic. You will get to the point where you can look at a question and go right to the answer. It will become scary.

Top Scoring Examination Strategies from Don McNea Fire School

1. -Read the directions very carefully and listen closely to the moderator or instructor if directions are given orally. If at any time you are unsure of any of the directions, raise your hand and a test monitor will come over and explain your question to you. These examinations differ from one section to the next. You should pay particular attention to the instructions for these types of examinations.

2. -Before you begin, make sure you have all the pages in the examination. In most examinations you will be told the number of pages in your booklet; check to make certain that you have all the pages or sections. If any page is missing, immediately raise your hand and inform the test monitor.

3.- Make sure that you are marking the right answer to the right question. All it takes is skipping one question and not skipping the corresponding number on the answer sheet, to cost you the examination. Every five questions or so, it is a good idea to take a look at the number in the test booklet and the number on your answer key to insure they match. Also pay strict attention to whether the answer key numbers are vertical or horizontal. You don't want to find out that you have been answering the questions on the wrong numbers.

4. -When marking your answers, make sure that you mark only one answer for each question. Do not make exceedingly large markings on your answer sheet; most of these examinations are graded by computer. If the marking is too close to another marking, it will be double keyed and you will lose credit for that question.

5. -If you need to erase an answer, be sure you erase it completely. Do not leave any shadows that could possibly show up when the computer is grading the examination.

6. -If you come across a question during the examination that you find difficult and you are spending too much time on it, skip over the question and leave a mark on your answer key. Do not mark in the area where you will be answering; mark to the left of the number so that you know to come back to this number. It is also a good idea, if you are allowed to mark in your test booklet, to mark out choices you have eliminated as being incorrect. This allows you, when you come back at the end of the test, to go back to only the choices remaining when you are seeking the best answer. If you come across a question on the examination that you find difficult, don't allow any more than two minutes on the question. If you don't know the answer, mark it, skip it, and return to it after you have completed the remainder of the test.

7. -Check the time during the examination. For example, if there is a 200-question test and a three-hour time limit, you should be on question 100 with 1-1/2 hours left. You should check the remaining time every 10–15 minutes to ensure you are on an appropriate time frame.

8. -Do not change answers unless you are absolutely positive. Time after time, studies have shown that when you change answers, 75–80% of the time you change it to a wrong answer. The only time you should change an answer is if you are absolutely positive or if you have miskeyed an answer. (For example, you intended to mark "C" and you inadvertently marked "B".)

9. -Don't be afraid to guess at an answer. Most firefighter examinations are scored based on the number of correct answers. On most examinations, there is no penalty for a wrong answer. If you have three minutes remaining on the examination and 15 questions to answer, try to answer as many as possible, but if time does not allow, at least put an answer down for every question.

Brent Collins is Assistant Fire Chief, Cleveland Fire Department and President of Don McNea Fire School. You can take a sample test with answers on our site: http://www.eatstress.com/donmcnea.htm

Reading Comprehension Test-Taking from Don McNea Fire School

Asst. Chief Brent Collins

When you are reading a short passage for the first time, read it carefully. A short passage is one that is only seven or eight lines long. You can retain all of the main ideas and remember where particular things are mentioned from one careful reading. Hence, you do not want to waste time reading this passage twice.

Besides wasting time, another bad consequence of reading a short passage very carelessly the first time is that it may leave you with some false impressions of what you have read. Wrong ideas can get stuck in your head from a careless reading. Then it will be more difficult to get the correct answer.

For long passages, look ahead to see what is being asked. Take a look at the "stem" of the question, the sentence that precedes the answer choices. And look at the kinds of choices that are being offered. Sometimes reading passages are long but the questions are asking only for particular details. In that case, you can often skim a long passage to find the particular detail.

Keep forging ahead. Do not get bogged down if there is a word or sentence you do not understand. You may get the main idea without knowing the individual word or sentence. Sometimes you can sense the meaning of the word from the context. Sometimes the word or sentence may not be the basis of any question. If there is some idea you need to answer a question but do not understand, read it one more time. If you still do not understand it, move on. You can come back to this question later if you have more time at the end of the test.

Picture what you read. Try to form a picture in your mind as you read. Schoolbooks used to teach reading contained many pictures because pictures aid comprehension. When reading material without pictures, it will aid your comprehension if you use your imagination to picture in your mind what you are reading. Read as if you were a professional illustrator who has been hired to do an illustration for the passage.

Ask yourself questions as you read. When you finish reading a sentence, ask yourself what the author was saying. At the end of a whole paragraph, ask yourself what the point of the whole paragraph was. If you ask yourself questions, you will find that you are paraphrasing the passage in your mind. That will help your understanding. Know where the author stands. Sometimes a passage will contain an evaluation of some ideas of tools or procedures. The author may want to make the point that certain practices or procedures are bad or that certain tools may not be right for a particular job. Be sure you know if the author is accepting or rejecting something.

Circle keywords and phrases. In a reading comprehension test you are not reading for just a vague general understanding of the passage. You usually have to read for detailed understanding. There will be individual words that are important for grasping a point exactly. You do not want to write so much on a passage that it is hard to read a second time if you need to go back to check a detail. But you do want to circle key words or phrases that will enable you to zero in on precise points needed to answer a question.

Brent Collins is Assistant Fire Chief, Cleveland Fire Department and President of Don McNea Fire School. You can take a sample test with answers on our site: http://www.eatstress.com/donmcnea.htm

Psych Questions on the Written

If you haven't already figured it out yet, there are psych questions being placed in the entry-level written test to eliminate candidates early in the process. The red-hot candidates, who have been the backbone of the fire service, are taken out by these purposeless questions. If you answer those questions as you normally would, you could fail that section. Go figure. Are you prepared? Most aren't. We've had several calls from candidates looking for a way to get over this hurdle. "Just because you're paranoid doesn't mean they're not after you."

The top program in getting ready for the written test is the Encyclopedia of Firefighter Examinations available on our web site. Several of our candidates have purchased this book. One of our candidates said if he had answered the psych questions on the LA County test like he normally would, he would have failed. He passed. His buddies didn't.

Here's a segment from The Don McNea Fire School Psychological Exam Preparation program:

Psychological Examination

In today's fire entry-level testing field, many departments are moving toward the direction of including psychological testing, not only during the last steps of the hiring process but during the initial testing phase. Some of these examinations are used to decrease the number of applicants taking the examination because of the cost involved in testing large numbers. When giving psychological tests as part of the written test, it will significantly reduce the number of applicants usually a minimum of 50%. The number one cause of failure on these types of examinations is not knowing what information these psychological tests are trying to retrieve from the fire applicant. The questions used on psychological examinations are determined by identifying major personality traits and characteristics of a successful firefighter. Knowing what personality traits and characteristics are being sought will improve your chance of scoring well on these examinations.

One fire applicant recently emailed us the following question:

One of the questions I am never sure how to answer on a psychological exam is whether a question should be answered "uncertain" or "not sure." There are questions that I have encountered during testing that I feel should be answered that way. My scores are not reflective of the individual I am and I was wondering if you have any insight into whether these questions could be making a difference in my score.

Answer: This is one of the most frequently asked questions from fire candidates taking psychological examinations. When you encounter a question that you aren't sure how to answer, we strongly urge you not to answer "uncertain" or "not sure." Doing this gives the impression that you are a fence-rider and not sure of decisions you make. In a psychological profile of a firefighter (on which many of these questions are based), they are looking for an individual who is very confident of the decisions he or she make.

Here's a testimony from a candidate who benefited from the Don McNea Fire School:

Thank you for helping me with the psych portion of the written exam. I took a test a year ago without knowing your information and scored in the low 70's. About 3 weeks ago I took that same test and scored a 94% and will be continuing in the process. That following week I took the same exam at my home department and though I do not know my score I have been invited to continue in their process as well.

I honestly believe that without knowing the information that you gave me and talking to you on the phone I would not be continuing in the process anywhere. I believe in what you have done for me and therefore I have recommended your material and website to many of the candidates that I have talked to in the past few weeks. I cannot say it enough but THANK YOU! THANK YOU!!! — Larry

Brent Collins is Assistant Fire Chief, Cleveland Fire. You can take a sample test with answers on our site: http://www.eatstress.com/donmcnea.htm

Take Any Test You Can!

You've got to be kidding me!

I've talked to three candidates recently who had the opportunity to take some killer entry-level tests. They didn't take these tests. When I asked why? They said, I don't want to work there. That would require me to move and my wife won't go, I'm waiting for the only department I want to work for now, or I only test in this region. Don't tell me how bad you want this job and then give me one of these stupid excuses.

I have several candidates who have gone out of their state to take tests in preparation for the "city they really want to work for". Guess what? They get offered jobs. How difficult would it be for you to turn a badge down? Guess where they live and work now? And, it's a lot easier to get a job once you have one. I know one candidate who went all the way to Wyoming to get his badge. Now he's testing back in his home state of Washington.

Understand the more tests you take, the better you will be at taking tests. Then, when the one you really want comes along, you're dialed up ready to nail that badge.

Just Take the Test

I believe in taking every test you can get to. Beg, borrow, grovel and do what ever it takes to get to as many tests as possible.

Cost? Spare me this part. This is an investment in a career position has a great salary with unbelievable benefits way into retirement.

Get serious. In case you haven't noticed, there is an army out there trying to get this job. You need every angle to make it happen for you. Firehire, CPS or what ever. Just get up, suit up and go do it. It will pay off biggggg in the future. You waste more time and money buying beer and pizza with your friends. Better odds? You bet. It could be your day. Your ticket into the fire service.

Life Can be Plan B

Jon and his 9-fire technology academy buddies set out to target six departments in the northwest they wanted to work for. Their plan A would cultivate these departments and be in a position when they tested. After almost two years no one got hired or was high enough to be considered. Then Jon read a section of our web site that encouraged candidates to test wherever they could get to.

This made sense to Jon especially when he figured out that he was only able to take around two tests a year. Like hands on academy and education skills if you don't use your oral board skills you will get rusty faster than trying to throw a 35' wood ladder or laying a line when you haven't don't it for awhile.

This is not taking into consideration that departments don't always test every two years, switch to medics only or hire only laterals.

So, non-medic Jon tried to convince his 9 buddies to expand their horizons and establish plan B to test any and every where they could to keep their oral board skills at the cutting edge. None of his buddies were interested because they believed that because of their academy training and education and how they were laying the ground work, it would only be a matter of time before one of the six departments on plan A would pay off.

In a short time non-medic Jon found out the more tests he took the better he got at taking tests. His oral board scores started climbing and he was getting called back for chief's interviews. Then BINGO! Jon gets a job offer from THE PREMIUM fire department in the southwest (yea, that one). As he was packing to leave he offered our program that helped him get hired to his buddies. He was surprised they weren't interested. Didn't need it. They were still banking on plan A.

It's now 3 years later and Jon's dream department, THE PREMIUM department in the state of Washington (yep, that's the one), announces their test. Guess what? Jon gets a job offer and gets to go home with his new bride, also from Washington.

Again he offers our program to his buddies. He is shocked again when they said they don't need it. So, how many of his 9 buddies were hired during this period of time? None, zip, nada.

Sometimes life can be plan B.

Don't be Late!

Two things stick in my mind at a Stockton, California written. One candidate drove all night from southern California with his buddies. When they went in, he realized he had forgotten his pictured driver's license. Did they let him take the test? No! He had to go wait outside for these friends to finish the test.

The second thing was a guy I saw running from two blocks away. He was really booting it. High stepping. He was trying to make it before they shut the doors. He wasn't a little bit late. But 20 minutes late. Did they let him in? Absolutely not! No matter what the reason.

I saw this happen at the Phoenix written. No less than 15 people were turned away at the door because they were too late. These folks were more than disappointed. Probably more so now with how many are still in the process. Yes, parking was a problem. But, that was the candidate's problem for not allowing enough time to anticipate the unexpected. I saw some candidates arriving in taxicabs. This was a creative resource. This is something they can use in the oral. As the training officer was turning the late arrivals away at the door he said, "If this is how you're going to be as a firefighter, we don't want you." Brutal? A tough lesson? But, he was right.

The point here is to check everything head to toe and allow more than enough time to get to the written site and relax before you go in. The doors at Phoenix opened at 7:00 A.M. I was with the first to arrive at 6:00 A.M. Here again there were hoards of candidates at a dead run trying to make it when the doors were being closed. It's a wonder someone didn't have a heart attack. The 02 was definitely needed. Wonder how many of these folks didn't make it through the PT.

Here's a "Nugget". When you get to a written, don't talk to anyone; not even your friends. Psych yourself up on the way. Separate yourself for game day. What would happen if someone gave you some news that could throw your written score off? Some news they had received results from a test you both took. That they were going forward in the process. And, you didn't get a letter. Troubling? Would this or any other information like this mess up your mind before a test? You bet.

This even applies at the oral. Orals can often run behind. Several candidates could be in the waiting area at the same time. Do not, I repeat, do not start a conversation. One of our candidates felt the nervous need to start talking to another candidate. He found out this guy was a scientist. He felt he was less and didn't know how he could compete with such overwhelming credentials. He blew his own game plan.

He regrouped on his next test and got a badge. He was probably more qualified than he gave himself credit for. Who knows what happened to the scientist?

Physical Agility

Often, candidates don't realize that it's not just strength in the physical agility. The "Nugget" is technique, momentum and grip. If you are uncertain or having problems in the physical, take advantage of any college or academy programs to learn the techniques to practice pulling hose, throwing a ladder, dragging a dummy (not you), etc. Many departments offer practice "run-through" sessions for their physical test prior to the actual date of testing. Don't pass up this opportunity.

You don't want any surprises during the physical agility. You need to have practiced hands on with every segment of the agility. Too many candidates think they are in great shape. One who did not take advantage of the practice session told me, "Hey, that 75 pound hose pack was heavy. Humping that hose bundle up the tower, hosting and other manipulative skills, then back down the tower steps made my lungs burn (they were still burning days later) and caused the loss of valuable seconds." The best way to train for this event is to up the cardio by going up and down bleachers with a backpack with weights or a weighted vest from http://www.weightvest.com

In those areas of concern, work with a trainer at a gym in those fields of motion that would improve your ability. Often fire training divisions know the exercises that would apply to those areas. When ice skaters were trying to break the record for a triple lux, they found working on upper body strength was the secret. You can learn more about physical agility training from http://eatstress.com/agility2.htm

Check in with your local area department and arrange to go by for a little coaching. What firefighter wouldn't want to puff out their chest showing his or her special techniques that got them their job or help on the fire ground? One of our candidates was losing sleep over the uncertainty of not being able to throw a ladder. These fears were put to rest after visiting a local fire department that showed the needed technique.

With ladder throws, it's gaining momentum and a continuous movement from beginning to end of the throw, using a pivot point and the weight of the ladder to your advantage. Dragging hose or a dummy is starting with a thrust to start up the momentum, taking shorter steps, keeping a low forward center of gravity, using your own weight to keep up the momentum during the pull.

Walking a ladder is using a pivot point and the weight of the ladder to your advantage. When raising the fly, pull the rope in short hand over hand movements in front of your face not much higher than your head. On each grip of the rope, turn your fist palm down to improve your grip. Keep one foot planted at the spur (bottom one side or the other) keep the other foot back for balance. Slightly tilt the ladder towards the wall for balance as you raise it.

The dummy from my son's department disappeared from the training center. Two days later a 911 call came in from a pay phone asking for help. When units arrived at the scene, here was the dummy standing up in the phone booth with the phone receiver to his ear. Case closed.

Many candidates feel if they set some kind of a record it will help in hiring. Not true! It is pass or fail. The secret "Nugget" here is to pace yourself. You don't have to break the record. If you would have no problem in passing the physical, then, why would you want to try and impress the training staff, the other candidates and tout that you set a new record? In your haste, you might injure yourself or fall down the stairs in the tower . . . and, you don't even pass. Now, you not only didn't pass the PT, you're out of the hiring process. How would you feel McFly?

Here are some valuable tips for CPAT from Tom Dominguez and Reed Norwood:

The secret to passing the CPAT is to be in shape with a high cardiovascular fitness level and to know the techniques as Captain Bob has mentioned. The average time is between nine minutes and ten minutes, twenty seconds. Try to think of the CPAT (or any agility) as a marathon where you are trying to complete the event instead of going for the record time. You can burn out if you are going for time no matter how well in shape you are.

Most people who fail the CPAT fail the first event (Stair Climb/Stair Stepper), or run out of time during the last event (Ceiling Breach). People who run out of time at the breach and pull lost a few seconds at all the prior event stations because they PAUSED to THINK of how to do the event or PAUSED or SLOWED down to catch their breath.

#1 Stair Climb: No matter how hard you train for the stair stepper, your legs are going to be like rubber after you get off the machine and start pulling hose. The recovery time for rubber legs depends on your fitness. Even still, rubber legs or not, you have to get moving and keep moving, and stay moving! If you stop at anytime during the events, the clock is ticking and you are losing time.

The tendency is that as you start wearing down on the stair stepper machine, your pace and stride will change and that will affect your balance. As you lose your balance, you start to wobble and the momentum of the weight on your body increases the swaying. As the distance of the sway increases, you will make a natural grab for the handrails. Grab the rail (more than twice?) to many times and you are disqualified. Instead of "grabbing the rail", use the back of your hand and push yourself back. Adjusting your stance and concentrating will help you avoid the "wobble." Just like wearing a SCBA, you also have to concentrate on your breathing.

#2 Hose Drag: As soon as you step off the stair machine, turn and face the line that takes you to the hose pull. As soon as the proctor takes the two sandbags off your shoulders, get moving! Pick up the nozzle and shoulder the hose and GO! This is not the time to worry about those rubber legs or try to catch your breath. MOVE! Go as fast as you can. Step into the box, turn around, get down on one knee (being careful not to come down too hard and injure your knee) and PULL the hose, hand-over-hand as fast as you can. That drum will give you some resistance when you turn the corner but if you're going at a good clip it won't be too difficult. You can breathe while hand-pulling the hose.

#3 Equipment Carry: When you get to the saw carry, just do it! Face the cabinet and remove each saw, one at a time. Now, turn around and pick up both saws. This will ensure that you have both saws touching the ground before you begin moving down the line.

#4 Ladder Raise and Extension: When you arrive at the ladder raise, get down, grab the rung and raise the ladder. You have to push the ladder up, rung-by-rung, as fast as you can. Move over to the fly extension and just do it.

#5 Forcible Entry: Breathe, as you follow the line and pick up the sledgehammer. Start swinging as soon as you can in short choppy strokes. Departments may set the forced entry device at a level that fits their needs. When the alarm sounds, let go of the sledgehammer and move to the tunnel crawl.

#6 Search: Get in and get out! You may not move like a greased pig at the fair but you do need to move. One candidate wrote: Here is where I lost about 15-20 seconds. The event itself is pretty fun if you are not claustrophobic. Be aware of the obstacles inside. I could not figure one out, and I got disoriented and lost precious time figuring it out. Crawl fast as there are no abrupt edges that you'll run into. All the walls are tapered so as long as you keep your head down you can fly through. Doing the practice "run-throughs" will take away all doubt of what and where the obstructions are in tunnel crawl.

Always remember to stay right, and come back to your right after an obstacle. The event is shaped in a horseshoe, so there are two right turns. This can be a good time to catch your breath as well in preparation for the dummy drag.

#7 Rescue: At the dummy pull, size up where the handles are before you get there. Grab them and get going. You may feel the burn in your legs but don't stop. It saps your strength to have to get the dummy moving again each time you stop. When you reach the barrel, do not make the turn until the dummy's knees are even with farthest side of the barrel. If you try to pull the dummy around the barrel any sooner, it takes more energy and it will take more time. Get over the line and let go of the dummy and get to the ceiling Breach and Pull.

#8 Ceiling Breach and Pull: This is the event where folks run out of time and fail the CPAT. Grab the pike pole and step in. Start pushing and pulling with all you got! If there's a D-handle on the pike pole put a hand under it for increased leverage. Get a rhythm/fast pace going. An object at rest requires energy to get it moving. An object that is moving requires less energy to keep it moving. If those ceiling hatches are not making lots of loud noise, you are not working very hard. You can buy yourself some time here that you may need to finish the CPAT in time.

Follow the instructions of the proctor! The proctor will either tell you where the line is or point to the line you are to follow. People have been failed for not following the right line to the next event.

If you were to pause five seconds at the start and stop of every event, or to stop and breathe or think about each event, you can loose about a minute and a half of precious time. Once this time is gone, you cannot get it back. This goes back to what Captain Bob was writing about when it comes to the manipulation and techniques of each event.

You can over-train by carrying extra weight in your backpack while you train for the stair stepper. Seventy-five pounds on your back places a tremendous amount of stress on your ankles, knees, hips and back. Practice the event as you are actually going to do it. Work out at the same pace and distance as the actual stair event. The stair stepper event (as are most of the CPAT events) is based on cardiovascular fitness and endurance. It is expected that you will be anaerobic and that is what the CPAT is attempting to do. While strength is required, you don't need to be an Olympic weight lifter.

Here are two link resources to gain information on the CPAT:
http://www.phoenix.gov/FIRE/recruit.html **and** http://firepat.mtsac.edu

Don't Get too Stressed Out

A candidate recently tested with a city for a firefighter position and had completed the written, skills test, and interview. The chief's oral is soon.

Here's what happened next: The top candidates participated in the physical agility. At the beginning of the physical the Chief stated that failing any portion of the physical will result in disqualification. We went in order of our scores. Number One went over the time limit in event #2. The Chief (present throughout the physical) turned to the rest of us and asked if it was OK for Number One to repeat the event. I remained silent while the other candidates mumbled OK. Number One repeated the event and was under the allotted time. Number Four was also unsuccessful on his first attempt at this event. He also was under time on his second attempt. The rest of the physical ran smoothly without anyone failing another portion.

At the conclusion of the interview I was going to speak to the chief about the fact that these candidates were allowed to continue in this process until he thanked us for the display of teamwork he witnessed today. I feel that this was not fair to the rest of the candidates to allow these guys to continue in the process. I feel for these guys but at the same time, I want a job. I was not allowed to try and better my written score or my interview score. Why should they get a second chance at the physical? Especially if the Chief stated in the beginning that failure of any portion of the physical would disqualify a candidate from the process.

This problem might resolve itself. Several firefighters were present at the physical and were obviously upset that the Chief allowed these candidates to continue. At any rate the chief's oral is this week. I wanted to know your opinion on this situation. What can or should I do? I want a job more than anything right now. I don't necessarily want to raise a beef with the chief because I know I will not win that battle. Please understand I feel like a bad guy for even having these thoughts. I have put in a lot of hard work to come this far and I feel I should now be higher on the list. What should I do?

Reply: I bet the chief wants the number one guy and had to let the number 4 retest to make it look fair.

Let it go! If you don't, it will turn you inside out and affect your chief's oral. The Chief is doing it his way. Had you talked to him, you would have never, ever been hired by his department. Remember you are the rookie. You don't have an opinion.

If you're going to start out before you get the job by fretting over this, how are you going to be in a 25-year career? These things have a way of working themselves out.

"Revenge is where I take the poison . . . and wait for you to die!"

Physical Stamina

It's tough enough to get this job. You need to be ready to keep it. I talked to another candidate who had a fire job. Then he lost it because he wasn't physically fit enough to make it through the academy. This is not the first time I have received this kind of call. What could these people be thinking? They just show up at the academy out of shape with no stamina.

Just because you passed the physical agility, doesn't mean you're ready for the academy. Departments know the standards for the physical agility have been lowered over the years. They're going to make sure you can do the job or they will wash you out. This is not a casual walk in the park where you can build up your strength, agility and stamina along the way. You need to hit the ground running. Often you will work harder in the academy than you will during your career.

Training officers have told me the problem is candidates might be able to complete one evolution in a CPAT or agility test but in a fire department academy the wheels start coming of the wagon when the have to do the same evolution two, three, four times or more and then move on to another evolution multiple times. Some candidates just haven't built up their endurance before they show up to complete the academy injury free.

You never want them to know where your goat is tied. Once the training staff determines you don't have what it takes, they will start documenting you. It's only a matter of time then. Even if you're able to squeak through training the whole department knows who and what you are by the time you show up for your first shift. Then, let the games begin. They will treat you like a rented mule.

If you get washed out, it's difficult but not impossible to get another job in the fire service. Especially if you're a medic.

How about adopting the program of being ready to go before you show up for the academy? We're not talking about just a diet here. Couch or computer mouse potatoes need not apply.

"More suicides have been committed by a knife and a fork than any other weapon"

Academy Preparation

I have completed the hiring process for a fire department. Since I am an older candidate, and work a full time job (approx. 60-75 hours per week) it doesn't leave me a lot of time for conditioning. Right now I am running 1.5 to 3 miles three to four days a week I also try to add strength training. So my question to you is; 1) At this pace do you think I will be ready for a March academy 2) If not what should I work on, and how much additional work should I be doing?

Do as much as necessary to get ready physically for the academy. This is no joke. I've seen more than a few candidates at all ages get washed out because they weren't ready. Yes, you need to get going on strength training. Ask the newest firefighters on your department what they went through in the academy and what they did to get ready. Make a posting on the firecarers.com asking this same question.

This is an investment in a career. You're going for all the marbles. Go to a gym and ask a personal trainer to get you started. You can learn more about physical agility training here: http://www.eatstress.com/agility2.htm

Attire

What Do I wear to a Job Interview?

I had a candidate tell me he went to an interview wearing a tie, suspenders and no jacket. I asked him, "Who did you think you were Larry King?" I asked him if they called him back for a Chief's interview? No. The defense rests.

The strongest non-verbal statement you can make in the oral board is what you wear. It is time to step up and make the investment. In the blink of an eye of those first few seconds referred to as the "Halo Effect", the panel is checking what you're wearing, eye contact, hand shake, choice of words, voice inflection and more.

Men: These are only suggestions. Do wear a wool suit in dark blue or gray. Pinstripes are fine, but avoid brown, black, or high fashion brightly colored suits. Black is a little too formal, more for dances, funerals and being a star in the movie Men in Black. If black is all you have, wear it.

Sport coats or blazers are out, so is polyester. Tie should be in a solid color such as navy, red, maroon, yellow stripe, or paisley print. Wear a white, off white, or pale blue long sleeved shirt in cotton or a cotton blend. Starch it no matter what the instructions say. No patterned shirts!

Understand you are applying for a snot nose rookie position. You have no time or rank with the department you are testing for. So don't wear your military, volunteer, other department, dogcatcher or other uniform to your interview.

A candidate from out of state called from Southern California the night before his oral board. He asked if it was all right if he wore his military uniform to his oral? I asked him if he had brought anything else to wear. He said no. He said his dad and other members of his family that are in law enforcement told him it would make him stand out. I told him to go ahead and wear his uniform and we would talk later. Did he get called back? No.

Don't: Wear casual or novelty watches, too much jewelry, monograms, religious, political, or fraternity affiliation accessories. Beards are out; mustaches are a gray area. When in doubt, shave it off. Don't wear cell phones or any other electronic leases.

When my son was trying to become a firefighter I begged him to shave off his mustache. He said Dad this took me 26-years to grow and I'm not going to shave it off. He got hired. He got married and his wife made him shave it off. Go figure.

Women: Do wear a tailored business-like suit or dress with a jacket, not overly feminine. Choose suits in conservative solid colors such as gray, navy blue, black, beige, or camel with conservative hemlines. Natural fibers such as wool are your best bets.

Don't: Wear anything flamboyant, trendy, faddish, low-cut, too tight or short, or otherwise provocative. You are not trying to make a fashion statement, but trying to get a badge! No heavy perfume, ankle bracelet, stockings with patterns, lace, bold colors, or seams; sandals, very high heels, unusual colors, or casual styles. Ladies: hair up; no bangs falling into your eyes or face. Don't ever wear slacks, even pantsuits.

> *From Anita: You say women should wear a business suit, but not a pantsuit to their oral interviews. What's the difference? During some coaching for interviews with a very 'oral board successful' captain friend of mine, I was told to never wear a dress to an interview because it makes women appear too feminine. So, I have always worn a black pantsuit to my interviews and people have always said that I have looked very professional. Is it hurting me? Thanks for your help! --Anita*

Reply: It's your choice, but I believe it's hurting you. You want to use everything to your advantage. When a women walks into an oral board wearing smart business attire, it changes the dimension of the interview trust me.

Yes, the panel members are supposed to be professional. Yeah, they're still men too. They can and will be distracted by what women look like, wear and how they interact.

If you don't think how you wear your hair, attire and accessories can be a distraction you are wrong. Being wrong could cost you the points you could have had going in prepared.

I was on a panel one day and a female candidate came in with long hair, a shorter skirt and a necklace that distracted us. Her long hair had been a little wind blown as she walked into the building.

We kept focusing on that pendant on her necklace. It looked like a naked women. It turned out to be a weight lifting woman flexing her muscles.

Like it or not, the suggest formula says: Ladies: hair up, no bangs falling into your eyes or face.

I had a female candidate contact me prior to an oral interview for the City of Oakland. This paramedic had been trying for 5 years to get on the fire department. She just missed the cut at Contra Costa County. She was tired of testing.

I asked her what she was going to wear. She said she always wore a pants suit. I convinced her it was time to step up and make the investment. She showed up for coaching in a tailored (killer) wool suit.

I showed her in 10 minutes on the video the mistakes she was making in her presentations.

She called me two weeks later on her birthday, that she had received her notice that she nailed that job in Oakland. She has the job of her dreams.

"Informal" Interview Attire

Although it can be called an "informal" interview, it's usually the chief's oral. It's called "informal" to get to know you better. But understand it's still for all the marbles. If you don't come across as the most enthusiastic candidate with the qualifications that meet this department's culture and objectives, you're out of the process.

The strongest non-verbal statement you can make is what you wear to any interview, especially the "informal" (chief's) interview. If your usual suit doesn't quite fit like it used to, it probably looks old and dated too. This alone can psych you out before you can say anything in the interview. It's time to step up and make the investment if you really want this job.

When my son Rob was going for his interviews, he had a new suit, shoes, tie, belt, socks and, yes, new underwear. He said it made him feel like the candidate he wanted to be.

Are simply slacks and a collared shirt OK?

NO! Be professional and look professional.

It doesn't have to be expensive. Go to the Men's Warehouse or a department store that has a sale. Ask for the personal shopper at many fine stores (this service is free). They will get you fixed up in your price range. It will make a big difference.

What if you have an oral interview coming up and they stated to come casual because if you pass you will go right outside and take the physical agility. What do you wear?

Reply: I know candidates who have been to this type of testing wearing sweats. How would you feel if you showed up in sweats and the majority of the other candidates had on suits (with a change of clothes to take the physical)? This has happened.

How about those interviews where you have been informed to wear casual attire?

Reply: Here again candidates have shown up and it looks like the high school prom. Consider wearing casual attire with a sports coat. If it is indeed casual, remove the sports coat.

Buttoned?

I am currently getting ready to graduate from a fire academy. The academy is conducting mock interviews for the cadets. I was considering going into the interview with a suit on, do I unbutton the jacket, leave the jacket buttoned, or remove the jacket and place it on the back of the chair?

Never, ever take your jacket off, even if it is 120 degrees and the panel has theirs off. How do you normally wear your jacket buttoned or not? Do it that way. You could go in with it buttoned and then unbutton your jacket after hand shakes but before you sit down. But, you might forget in the excitement of the moment. So, go in the way you would feel most comfortable.

Tom had a hand crew test where the interviews were held within minutes after completion of the agility. The agency expected dripping wet and smelly candidates to interview. He took along a suit and a few towels and found a hose bib and washed up as best he could.

Tom put his suit on and went into the interview. As it turned out, the interview panel was surprised that someone would wear a suit for a hand crew job. Needless to say, he got the job offer!

Tom said, "From time to time, I take classes held at various fire agencies. When those agencies are holding interviews and I am taking a class, during breaks, I still see numerous candidates showing up in Levis, Dockers and shirts with no ties. No point in shooting yourself in the foot! The moral of the story is ALWAYS wear a suit! Wear a suit for any interview, no matter what the title of the position, unless you have been instructed otherwise."

> *There's nothing quite like the look on a candidate's face when he enters the "holding pen" room where everyone waits to be called for their turn to interview, and realizes that he's the only one not wearing a suit.*
>
> *Don't be that guy.*
>
> *I was "that guy"!! I prepared so hard for my interview that I totally forgot to ask people what to wear. I just assumed slacks and a dress shirt would work. Wearing a suit never even crossed my mind!*
>
> *Turns out, I was the only one in the waiting room not wearing a suit!! Everyone else was so GQ that I felt like I was sitting in a room full of investment bankers. I could tell everyone was looking at me like there was one less candidate to give him or her competition.*
>
> *To make matters worse, my interview was rocky, and one of the captains proceeded to grill me relentlessly, and guffawed at all my responses. In the following two weeks I was convinced I failed. But I got the call in the end! I start soon.*
>
> *I wouldn't recommend that to anyone. It was definitely a psychological disadvantage to walk into a room full of slick candidates and then have to walk into the interview with an air of confidence. It was even worse waiting for the results thinking that I blew it by not wearing a coat and tie.*

Reply: Candidates will tell me that I don't have a suit or the money to by a nice suit for my oral board. My advice. Rent one and look and feel like the professional you want to be.

During a coaching session one day a guy told me whenever he begins the testing process with a city he (yes, this was a guy) goes to Victoria Secrets and buys special underwear. He wears this special underwear for each step of the testing process from written through the oral for that particular department. I didn't need to know that.

The Oral Interview

According to retired Battalion Chief Dennis O'Sullivan, "The oral interview gets you the job! This is where you putt for dollars. Understand one very important thing here. If I'm on you're interview panel and your my kind of guy or gal, I will fill in some of the blanks to make up for your short comings. If you start off without establishing this natural bridge and being a know-it-all ass, I will never fill in any of your shortcomings. This is human nature."

This became crystal clear at an interview when a lateral candidate was asked how he would throw a ladder. After explaining how he would remove the ladder from the rig and the procedures to raise it, he went blank when they asked him where he would place the ladder.

Then, one of the panel members said, "Little Jack Horner sat . . ." The candidate smiled and remembered that ladders are placed in the corner of the building.

Oral Board Preparation

I Suck

"I suck" . . . I said this to myself after the first mock interview, and every time I prepare for an interview. I'm so glad it wasn't the real thing.

Are ready for your interview?

Many people we coach feel over prepared during their oral interview. They kept waiting for the other shoe to drop. It never did. One guy said "I felt like I was ready to dig a post hole with a back hoe".

How prepared are you? Have you figured out that you suck and started working to get to the point where you are over prepared? Don't fall into the trap of figuring it will take you a week to prepare, and then count back from the date you THINK your oral will be.

Most of the calls I get for coaching are from people who are so surprised that their interview is suddenly upon them, and they thought they had at least another week or so. It takes at least a week to memorize a script to the point it sounds good.

Do yourself a favor, sometime this week go to a mock interview, talk into a hand-held voice recorder, find out if you also suck, and get to work.

Don't be the guy who finds out he sucks in his interview, and makes the whole oral board watch you die the slow death. Good Luck and be prepared.

Captain Rob web site: http://www.myfireinterview.com   e-mail: mailto:captrob@sonic.net

Are you Prepared?

One of the guys I work out with at the gym with has a son who has been trying to get a fire job. He has all the usual credentials. Firefighter 1, almost a BA, 3 seasons with Cal Fire, yada, yada, yada.

He has been testing for 5 years. His dad gave him a coaching session just prior to his oral for his dream department. Dave had been practicing with a voice recorder. During the coaching session, Dave expressed his burning desire, passion, "my life won't be complete until I get a badge" compassionate longing, agonizing story.

One problem. Dave sucked big time! Even after testing for 5 years, he wasn't ready for any oral board. His answers were garbage. This should be no surprise, because most candidates are not ready either. Coaching usually takes about an hour. We ended at 2 hours. His closing was a dog and pony show (I wished this candidate would just end and get out of the room) pathetic mess.

I asked Dave how he expected to get a badge when he hadn't spent the time to get ready for an oral. He said, like most candidates, (a big clue here), he thought he was. This is what most candidates think. Does this sound like you? SFFD Captain Bill Long was a rater on a recent oral board. He said you knew which candidates were really prepared. Those prepared candidates caused you to straighten up in your chair.

The important point to realize is it doesn't take much to improve your situation and separate yourself from the clone candidates. Dave only had a couple of days to review his coaching tape and redial his approach.

He called me the day after his interview. He sounded like he didn't step on any land mines, wasn't stumped and was able to put it together to make a real good presentation.

A few days later, there was a message on my recorder. A guy was yelling, Captain Bob, you are the man. It was Dave. He had just received the call that he was going to the Chief's Oral. His first in the five years he had been testing. Not only was he going to the Chief's Oral. He was number . . . 2! They were interviewing 30 candidates for 10 jobs. How do you like those odds?

When you are going for all the marbles, don't you want an insurance policy to insure your chances for getting that badge?

Wouldn't you want someone to tell you that you sucked big time before you head into your next oral? There's an oral board in your future. Do you want to be telling yourself 'I suck' coming out of your next oral and you will do better next time? Or, have that feeling that you knew 'I smoked it' and it was going to get you that badge?" When you're ready to come in from the wilderness, drop your ego and get to work.

Candidates just like you are trying to send the message that this program works and it can happen quicker than you think.

Paul said it best:

I didn't realize how incredibly dead in the water I was until I realized that what I thought was unique in my oral boards was truly another fine example of a clone candidate.

Sadly, I know that there are thousands of other candidates shooting themselves in the foot, being difficult on themselves, telling themselves that they aren't cut out for the job because they've tested so many places and keep getting low on the list . . . or not getting on the list at all.

I FINALLY got the job in a busy, full-time mid-western town. All of my dreams have instantly come true. — Paul

Speaking Skills

Battalion Chief Steve Lewis said, "If you are a good public speaker I can't guarantee you will get a job in the fire service, but if you're not a good public speaker, I can guarantee you'll never get a job as a firefighter".

At El Camino College Battalion Chief Steve Lewis, asked the audience, "Who here works out?" Everyone raised their hands. Then he addressed a really fit guy in the front of the audience and asked, "How often do you work out?" The reply, "Two hours a day". That was clear. The guy was in good shape. Then he asked the guy, "How often do you practice your interview skills?" The reply, blank stare.

I thought, "Wow, good point. I never thought of it that way." Most of us have youth on our side and could smoke the agility without any preparation, but the agility is just pass/fail. So is the written. The interview is generally 100% of your score. GOOD LUCK! Chris Bertrand

Sage advice. Speaking skills should be the cornerstone of every firefighter candidate who is pursuing a badge. Getting a firefighter job is all presentation skills!

Interesting enough speaking in front of a group is the top rated fear. The second greatest fear is dying. A guy was speaking at the funeral of his friend. He said, "My friend laying in the casket is better off than I am up here speaking." Those who have been panel members on oral boards have first hand knowledge. We see candidates with great credentials but don't have the speaking skills to present them. To improve your skills, take a public-speaking class at a local college, get coaching (check out the Coaching session page), or join Toastmasters.

Sample Thirty Plus Oral Board Questions

I believe there are only about 30 oral board questions. Plus or minus a couple. But these 30 can be disguised into hundreds of different questions.

1. Tell us about yourself.

2. Why do you want to be a firefighter? When did you decide on this career?

3. What is the job of a firefighter? Are you qualified?

4. What have you done to prepare for this position?

5. What are you bringing to the job?

6. Why do you want to work for this city or agency?

7. What do you know about his city or agency?

8. What do you like to do? What are your hobbies?

9. What are your strengths? Weaknesses?

10. What would your employer say about you?

11. What are the attributes of a firefighter? What is the most important one to you?

How Would You Handle the Following Scenarios:

12. Drinking or drugs on the job? 13. Stealing on the job? 14. Conflict with another employee?

15. Irate citizen? 16. An employee crisis at an emergency? 17. Sexual harassment?

18. Racial situation? 19. Conflicting orders at an emergency? 20. An order that could place you in great danger or be morally wrong?

21. What do you say when you don't know an answer to a question?

22. Are you on any other hiring lists? What would you do if another city called you?

23. When can you start if we offered you the job?

24. How far do you want to go in the fire service? Where do you see yourself in 5 years?

25. What are the quality traits of a firefighter? Which one is the most important to you?

26. Have you ever been in an emergency situation? Tell us what you did?

27. What word would best describe you in a positive way? A negative way?

28. How do you handle conflict?

29. Why would we select you over the other candidates?

30. Do you have anything to add?

It's your job to take off the disguise and find the real question and have a "Nugget" answer to satisfy the oral board, get your best score on the answer and cause the board to go onto the next question. This "Nugget" tool is one of several that can separate you from number 40 and below on a list and put you between 1 and 10 where you get a shot at the badge. You'll know the difference when the call comes in to go to the Chief's oral.

Five Nuggets for a
Successful Job Interview

Simple Tools to Uncomplicate the Process

1. -The job interview is like auditioning for a play. You must know your lines for the part. Do you meet the minimum requirements?

2. -To learn your part, make an outline of why you want this position, what you have done to prepare, why you want to work for this agency, etc. It must be about you; not a clone of someone else.

3. -The outline will become your script to audition for the part. Practice, practice, rehearse, rehearse, and over-learn the part until it becomes second nature to you. This will help prevent stage fright.

4. -With tremendous enthusiasm, use your new role to capture the first 32 seconds of your audition. This creates its own energy.

5. -Don't reiterate in your closing. Use only the key points not already covered in your script. Without being boring, tell the interviewers that you really want the job and with your qualifications hope to be considered for the position. Make a cordial ending. Then, shut up and get out of the building.

There are Six Steps
To Answering Oral Board Questions

You should have a script (sample here: http://www.eatstress.com/workboolette.htm) that you have rehearsed with a voice recorder of anticipated questions by the time of your oral board. At the interview use these six simple steps in answering oral board questions:

1. -Actively listen to the entire question. I have seen candidates stop listening when they think they already have the answer. They don't. Listen!

2. -Make sure you understand the question. If not, have the question repeated or rephrased.

3. -Pause and gather your thoughts. It might seem like an eternity, but pausing is an acceptable tactic to show interviewers you are paying attention. During the pause, you can figure out the root of what they are asking.

4. -Ask the question or make the statement to clarify the question. The question might be, "You see your partner pick up something at an emergency scene, what are you going to do?" Taking the question down to its basic form, what is the issue? Stealing. Then, formulate a simple answer. For example, you might say, "I would ask, 'Is that yours?'" The board is going to tell you that he is taking it, but you already scored the points. After asking the question, you determine your partner is stealing, what do you do? Since stealing is an ethical issue and he won't put it back, you might say, "Why don't we go to our supervisor?" Why? Stealing is against the law.

5. -KISS. Keep it simple sweetie. Don't start a soap opera. Most candidates complicate the process. They intellectualize their answers, run past the question, decide an answer before hearing the entire question and fail to understand the process.

My son, Rob, was interviewing and the question was asked, "You have just finished your interview and go outside and find a man down on the sidewalk. What would you do?" He answered, "I would go up and say, 'Buddy, are you all right?'" Someone finally got the answer right. For three days, job candidates were saying things like "Activate the 9-1-1 system" and "I know CPR."

6. -Deliver the Nugget answer with enthusiasm! Your personalized Nugget answer will set you apart from the clones.

A word to women. You have the advantage of bringing more feelings and emotions to your answers at an interview. Be careful though. I've had women at interviews start talking and it was like going on a journey. There seemed to be no final destination. Most men on the panel were not packed for the trip. This can't be like a conversation with your girl friend. You have about 20 minutes to give complete but concise answers.

Getting this job is a process. Here is a testimony how quickly it can happen:

Subject: Another success story

I am writing to thank you again for your entry level program. I sent you an e-mail after my first interview outlining how much your program had improved my presentation. At that time, I mentioned that decisions were not expected until July.

To my surprise, 8 days later I received a call from personnel. Could I attend a Chief's Oral next week!! I was 1 of 30 to be called in for 23 jobs (The first interviews involved 900 candidates). My first thoughts were "stay on that winning pony".

After the Chief's Oral, the good news kept coming. I got the call two days later with a conditional offer of employment! The background and medicals were done the following week.

My most recent phone call came yesterday. I heard the words I have been working towards for six years- Congratulations, you have been accepted as a probationary fire fighter!! Uniform and equipment sizing is set for Saturday, can you attend?

I can't stress enough how much your program helped me. I will be sure to recommend you to anyone I can. Thanks again! — Brian

Oral Boards

What are you actually doing going to an oral board? If you answered: selling yourself, making a good impression, and, yes, don't forget to ask for the job. But, what you're really doing is auditioning for the part to be a firefighter, engineer, inspector or officer. Just like the part in a play. Do you know your lines? Do you know your part? If you went down to a local college to audition for a part in the community play, you have to know your part and lines wouldn't you? Right? It's the same thing in an oral board. You have to know what you're going to say before you sit in the chair.

Does a Broadway play start on Broadway? Of course not. It starts in Iowa, Miami or Connecticut. They take it on the road to try it out, work out the script, refine and polish it up. If they create enough interest, sell enough tickets and get great reviews from the critics, they make it to the bright lights of Broadway. It's the same in getting ready for your oral boards. You have to take this on the road to get ready for your oral boards. You have to get your script down. A script about you, not a clone of someone else. Then, you practice, practice, practice. Rehearse, rehearse, rehearse until it becomes second nature to you. Once you do this it will be in your subconscious. That's where the magic takes place.

Here's how it works:

I recently tested for the City of Denver. The written test was the first step in the process and over 2,100 people showed for the test. As I waited outside the building, I thought to myself, "Do I really have a chance at getting a job here. I have worked in the fire service for the past eight years (7 years as a military firefighter and one year paid in NM)."

This was going to be a challenge for myself. I received the results for the written and I passed. Come to find out 1,400 people failed that test. They took 757 people to the oral boards. I didn't fare so well in Colorado Springs oral boards last year. I ranked something like 173 out of 250. After being in the fire service for eight years, I thought I knew everything needed to be hired.

I studied your program, used a voice recorder, and practiced, practiced, practiced! I found out quickly what I didn't know. I went to the oral boards prepared. I as I walked in they stated 757 people were going through the boards. I tried to remain positive and just present my package. I was in and out of the interview quickly. I walked to the car and my wife asked how it went. I said, good but I'll find out in two weeks.

Yesterday, I went to the post office to get my mail and there was the letter from the City of Denver. I was too nervous to open it. Finally I decided it was time, my rank was 14th out of 757. I couldn't believe it. Fourteenth. WOW my total score on the board was 100.0000%. I ACED IT!!!!!! I immediately called my wife at work with the good news. She cried. Thanks Bob. — Jason

Voice Recorder — The Miracle Oral Board Tool

I can't believe how many candidates I talk to in person and on the phone that don't know how they sound to others. They're soft spoken, monotone like they're giving a patient assessment, with no enthusiasm or inflection with a lot of pause fillers. Many are not using a simple tool that could tilt things in their favor.

A recent candidate had such a monotone voice that I asked if he knew. He said, "Yea, but that's just my voice". I told him I didn't believe that for a second. What can I do about it? I've been testing where I can for four years, going to school and work as a federal firefighter.

Trying to get on his turf, I asked him during his coaching session what do you do with your time off? What is your interest, hobbies? What really rings your bell? Nothing seemed to work to break his monotone voice.

That was until a few days later I get a call from an energized candidate. I didn't recognize the voice. Yes, it was Mr. Monotone. He told me he didn't realize how bad it was until he listened to the recording of his coaching session. He said, "Man I sounded retarded. I can't believe how much stuff I left out. How many times I said "What Ever" and other stupid pause fillers I didn't know I was using."

The mystery of why this super qualified candidate could not get hired was solved by listening to a voice recording of what the panel had been hearing for four years.

So, what tools can you use to practice and rehearse your oral board answers? A video camera? Sure. You need to see how you look in action. But you are trapped with a video camera. A Mirror? Sure standing in front of a mirror is good. But you are missing the most valuable tool of all. A hand-held voice recorder that goes everywhere your car keys go. The closest distance between you and the badge is picking up a recorder and hearing what's coming out of your mouth like Mr. Monotone!

Few candidates have a script that they have been religiously practicing with a voice recorder. Ninety-nine percent of the candidates I ask aren't. I asked a college program recently how many had been practicing with a voice recorder daily? No hands. How about weekly then? Nope, None. O.K. how about monthly? Finally three hands went up out of a total of 40. Then, like Mr. Monotone don't be confused by why you're not getting high enough on the list to get a call back to play the part of a firefighter. The mystery has been solved.

Many applicants want this job so bad they will do almost anything ethically and morally to get it. I guess that doesn't include using a voice recorder to get your timing, inflection, volume, where to cut out material, get rid of the uhs and other pause fillers, or to find out if you really sound like Donald Duck. You need to get married to your hand-held voice recorder. You need to hear what the oral board is going to hear out of your mouth. It's narrows the distance between you and the badge you're looking for!

This is usually a guy thing. Guys think about their answers in their head and write them down. Then they think their answers are going to come out of their mouths like magic in the oral. Trust me, they don't! The brain and mouth don't work that way.

Try this. Take 3X5 cards and write down your oral board questions. You can get a sample list here: http://www.eatstress.com/thirty22.htm Practice your answers with a hand-held recorder that goes everywhere your car keys go. If you hear something you do not like when you play it back, turn over the 3X5 card and write it down. The next time you go after that question, turn over the card first and see what you don't want to say.

Let me tell you how critical this really is. If you're not using a voice recorder to practice, practice, practice, rehearse, rehearse, rehearse and over learn your material until it becomes second nature to you, you might as well not show up for the interview. You are wasting the oral board's time and your time! Seek out another career. Understand you still have to interview there too. The above candidate has already lost some great opportunities. Had he been faithfully using a voice recorder to prepare for his oral boards, he probably could have had a badge already.

Some will say, "Well, if I practice it too much it will sound canned." NO it won't! It sure will be planned though. Practice makes permanent. "Luck is preparation meeting opportunity." One practice session with a tape recorder is worth 10 speaking outlouds. After practicing, you will get to a point where your answers will get into your subconscious. That's where the magic begins. You can't be fooled.

We think practicing with a voice recorder is so important; we will not do private coaching with a candidate if they aren't using one. It is a waste of our time and their money. Be advised that your competition knows the value of using a voice recorder. They are catapulting past you if you're not using one too.

FireTeam Video Testing

Many agencies are being convinced to use the new type of video testing because they can test large groups at once at a lower cost and eliminate candidates early on in the process with psychological video scenarios.

Candidates who have trouble with the FireTeam Testing video found that on the difficult questions if they went to their feminine side they answered better. Isn't that precious?

This from a candidate who paid to test drive the video testing:

I went to the website www.fireteamtest.com and paid a fee to take the practice test. The test takes 30 to 45 minutes so unplug your phone and listen to the instructions VERY carefully; my phone rang while I was watching the video and missed an important part. DON'T answer the situational questions the way you think the "test writers" want you to, answer thinking about the big picture and what is right for everyone, being a kiss-up will not get you a 100% on this test. I know; I got kinda toady on one question and got slammed. So much for my 100%

It seemed like for each question the four answers fell into these categories:

Correct
Semi-Correct
Neutral
Wrong

Of course if you sit there and try to figure out which ones fall into each category you will fall behind and start screwing up.

Check it out at www.fireteamtest.com

Fantasyland

The oral interview is like fantasyland. It is not like the real world.

What is the biggest problem candidates have going to an oral board? Fear, nervousness, and anxiety? What causes it? It comes from feeling you are not prepared. It's not uncommon to open the door to call in the next candidate at an oral board and as they start to stand they freeze up like the Tin Man in the *Wizard of Oz*. They have sweaty palms when shaking hands. This is stage fright in auditioning for your part in the play to become a firefighter. This is not how you want to show up at your oral boards.

As I mentioned, everyone has butterflies. The trick is to get all the butterflies to fly in the same formation. Practicing with a voice recorder will remove up to 75% of the butterflies. You will need the other 25% to carry you through.

When you see an actor in a play or a movie, do you see the actor or the part they are playing? Sure you see them in the part. That's because they have practiced and over learned the part until it became second nature to them. They were comfortable playing the role. You can too, by taking as many tests as you can. The more tests (auditions and performances) you take, the more comfortable you get at taking tests. So, when the test for the department you really want to work for comes along you are up to speed, ahead of the curve to nail that badge!

The oral interview is like fantasyland. Often not like the real world. Your answers in the oral board might not be what you would do in real life. Don't fall into the trap. The board understands the rules, you can't fool them. If you try, the board will crank up the music and let you dance your fool head off. Don't try to intellectualize or bring heavy logic to this process.

My son Rob's captain was on an oral board panel for 5 days. One candidate got the top scores from all 3-panel members. As they finished with this candidate and were totaling their scores, the candidate said, "I'm sure glad that's over." The panel members acknowledged. He said, "Because they're coming." A panel member asked, "They're coming? Who is coming?" The candidate said, "The Martians are coming." The panel members laughed. The candidate got mad. He was serious. He was a genuine kook. The reason he did so well in the oral board is he lived in fantasyland and he under stood the rules. This is a true story. Don't forget it. If you do, someone (maybe the village idiot), who understands the rules in fantasyland will get your badge. So, please follow the yellow brick road rules in fantasyland and don't look behind the curtain.

Don't be A Clone Candidate!

It's not the interview questions that are the problem, it's the answers! Unfortunately many candidates become clones and give clone answers. And the bigger problem is they don't know it. I hate to say, but often they are cloned in fire colleges and academies. Clone answers can doom your oral board.

One of our officers was on an oral board for a big city. Several boards interviewed 965 candidates. His board interviewed candidates over a period of 10 days. Imagine you were this officer and it is the fifth day of interviewing. You have just come back from lunch where the city has wined and dined you. You're tired and you know you have another five days of interviews ahead of you.

The next candidate is called in. The first question you ask is, "What sparked your interest and why do you want to be a firefighter?" He proceeds to give you the same clone answers you have heard from almost every candidate for five days. Public service, helping people, not the same thing every day, blah, blah, blah. The magic that you needed to hook up with the oral board has passed and you didn't hook them into listening to your stuff. You have just scored yourself. Trust me. You can see the glaze come over the raters' eyes. It's like a deer caught in the headlights. They are gone and they won't come back.

It's not that you can't use clone answers. You can. But first you need to deliver a signature story about you. Not a clone answer of anyone else. I haven't met a candidate yet that couldn't come up with signature stories. Signature stories demonstrate experience. They also tell that you not only know the answer to a question, you've lived it. Firefighters love firefighter stories. If you open up with a signature story, you instantly separate yourself from the other clone candidates. Stories show the oral board who you really are. You capture the board and take them on a journey with a story they have never heard. Is this making sense?

The toughest thing for candidates to do in an oral is being themselves on purpose. When you are yourself, you become conversational because you are on your own turf. This alone can lower the stress and the butterflies.

An oral board member told me they had a candidate who didn't answer all the questions the way they wanted him to do, but he had such great personal life experience in his answers (stories), they hired him anyway. This is human nature. Stories help bridge that gap. Clone answers and clone candidates don't have a chance here.

Example:

I was doing private coaching session with a candidate. He was telling a story about being a federal firefighter in Yellowstone when it burned. The story was not too exciting the way he was telling it. I had to stop and ask, "It sounds like you were trapped?" He was. Now he tells that story and the hairs start standing up on the back of your neck. You're trapped with him. You can smell the smoke and see the embers dropping around you. Does this story make a difference? Please say yes.

So the point here is not the question, but the answer. Start establishing your personalized stories.

Instant "Clones"

Recently I had the opportunity to participate in mock orals with one of my instructors who happens to be really great when it comes interviewing. In our class that comprises mostly of people starting fire tech classes, nobody did very well. It was a great lesson about how we need to start preparing and getting to familiarize ourselves with the testing process. However, 2 guys who were friends with our instructor participated in our mock orals, and put the rest of us to shame. They obviously have spent countless hours practicing orals with our instructor.

They really knew their stuff and not having any oral experience myself, I was very impressed, along with the rest of my class. My question is that these guys were so well rehearsed and knew each question and answers like the back of their hand, they sounded like actors in a play—anybody could tell that everything down to expressions, and hand motions had been practiced over and over to perfection. Is this what interviewers want when they interview you? Do they really want to see rehearsed answers? Don't get me wrong, the answers were very good, but seemed so artificial. Please let me know if it's better to answer questions to the best of your knowledge, or just to memorize good answers. Thanks, any input would be great.

What you saw was a perfect example of turning candidates into Clones. It's impressive at first. But if you felt is was too rehearsed, so will the oral board panel. When you see it over and over again it gets old and puts the panel into a daze. We could tell who the instructors were on many of the clone candidates by the second question. This will stick out in an interview. One thing about clone candidates; they will end up with a score that will put them in the clone pack.

One of our officers was going to be on an oral board panel for our department. He had been telling people that he could tell which candidates we coached. After the interviews, he was telling us about this great candidate who nailed his interview and came out number one. I asked him if he thought the guy had been coached. He said he was so good using his own stuff he couldn't have been.

When I told him this was one of my candidates, he screamed . . . NO WAY! Yep, he's one of our guys. Not only that, this guy had been testing for over 3 years. He scored 532 on his last test in Stockton. He came to us three weeks before his oral with our department. He had great stuff, but didn't know how to present it.

The proof is in the badge, and, as you already know . . . Nothing counts 'til you have the badge . . . Nothing!"

Who Are They?

I got off the phone with a candidate looking for advice. I asked him what his approach was to oral boards. He said I've tried to be open and honest to let them know who I am. Who told you to do that? A friend who is a firefighter. Well, how are you ending up on the list? Ted said, "I haven't passed any."

That's my point. Everyone becomes an expert once they've taken a few orals, has a friend who is a firefighter or someone tells them, "They said this is the way to do it". I marvel.

During their oral board clouds start coming out of their mouths. It's like hearing a soap opera. Yes, be informed. But a great opportunity is being missed here by not focusing on who you really are. Yes, I've seen the web site postings like, "I smoked that oral, or I kicked butt on that oral." But, don't touch that dial. I hope those folks will get back to us when the final scores are made available. And, get back to us on the day when your head drops after reading the letter that you didn't make it again. I hope you do make it. God only knows how much you have invested to deserve the job.

This from Fred:

You said, (not they said) "If I concentrated on my own stuff, used a voice recorder to practice and did the private coaching miracles could happen. It did in a big way. Faster than I thought. Not only was I hired; I was number one of the five they hired. Imagine me number one! An early Christmas present. Eternally yours, — Fred

The defense rests!

Using Humor in an Interview

Unless you're a humorous person, don't plan on laying something funny on the panel. I've seen people that weren't funny to begin with try to include humor in a presentation. It bombed. How would you feel in that situation if the room went dead silent and everyone just stared at you? What if this humor was in your opening statement?

This happened to Jim. He said it threw off his timing and confidence and he really never recovered. If something funny happens naturally during your interview or presentation, that's a gift. Don't plan on it happening.

Not a Regular Job Interview!

You're looking for a seamless no surprises interview in trying to obtain a badge. Stumbling just one question or a question you have never heard before could be enough to keep you from making the cut and going forward in the hiring process.

Many regular job and corporate interview candidates are stunned and baffled when they don't have high scores on their firefighter interviews like this one from Jeff:

Captain Bob: I just received my oral board score. The score did not represent how I felt I did during the interview. This is a big problem for me because I now realize that I DON'T KNOW what the board was looking for. I make presentations for a living, so I felt confident in what I did to prepare. I was sure that I just about nailed it. I've always been competitive about what I set out to accomplish, using every tool that I can utilize to reach my goal.

I've been preparing for these orals for months and feel extremely prepared. I don't want to waste another oral board without knowing that I've done all that I can to be the best candidate possible. Jeff

Reply: Jeff, You're not alone here. You have discovered like many other's that a fire department oral board is different than anything you have encountered. Too many candidates have beaten their heads against the wall for years getting to the point where you are now.

This could help to explain why:

After my interview rejection at an east coast city last week, I sent a letter to the D/C thanking him for the opportunity and telling him I'd appreciate any feedback from the interview. Well - he was honest - he indicated he wanted me to keep testing & interviewing, but wrote that I:

-talked too much & over answered the questions

- talked too fast

- some of my answers were based on book knowledge (?)

Also - as I mentioned I approached this like I would a corporate interview (BAD IDEA) and I tried to 'close' them at the end - they asked if I had any closing questions (jeez I wish I read your web site before going in) and I opened my big stupid mouth to say 'I kept your rejection letter from last year (I actually showed it to them - this was my second time interviewing) and it mentioned that candidates had failed to prepare and properly sell themselves to the board. I've been working to improve myself in these areas for the past year - have I properly done this?'

The D/C mentioned in his letter back to me that, 'I don't think that showing the letter I mailed out last time was the best thing you could have done. It was as if you were showing it off and showing us that you still had it.' I need to keep my damn mouth shut and just answer the questions. Live and learn. Stay safe Dave

What's your Competitive Advantage?

Every Saturday there's a farmer's market in our town. Many booths sell strawberries. There is one strawberry booth though that has a long line until they sell out. If you asked anyone in the long line they would tell you that these strawberries are the best. People stand in line for up to 20 minutes. What makes these strawberries better?

Well, it is known that once strawberries are picked, they will not get any sweeter. So this farmer picks his berries at the peak of their sweetness. One or two days later these berries would go to mush. But, this farmer knows he can pick his berries at their peak and have them sold the next morning. He also piles on extra berries as he bags up your purchase. This is his competitive advantage.

What is your uniqueness and competitive advantage in getting a badge? Something that is going to make you stand out among the other (strawberry booth) candidates? Is it your special uniqueness to take a program from inception to completion? Your long trail of customer service? Team concept? Strengths? The burning desire to continue when others would quit?

Your special uniqueness doesn't have to be firefighter related. As a matter of fact, those candidates that can weave their personal life experience into their oral board answers run right by the other clone candidates.

The point here is start writing down your special unique qualities that will give you that competitive advantage on your oral board game day. These will inspire the board to say we want to hire this person. It can happen just that fast when you know what it is for you.

Stories Get Badges!

Don Hewitt, one of the pioneers of television news and the creator of CBS's "60 Minutes" said, "The key to my success is four words that every child in the world knows. Tell me a story. Learn how to tell a story and you will be a success."

We encourage candidates to lace their answers with stories of their personal life experiences. Since no one else can tell a candidate's life experience stories they can't be placed in the mold of a profile. They become unique, fresh and convincing. In a recent fire academy all but two recruits were candidates who went through our program. You couldn't tell one from the other in the oral board because they were using their own stuff. Not a profile robot "clone" of everyone else.

If you have all the education, experience and the burning desire to get that badge and you're not getting hired, having to cool your heels in another position waiting for that next opportunity (not a bad ideal), you have be asking yourself why?

You can talk all you want about what we do here, how you want it or think it should be, but the candidates you are reading about in our material are a lot like you. They simply got positive results by putting simple techniques into action. The big difference is they figured out how to maximize the points in their oral boards, are now riding big red and taking home a paycheck.

Here's how they did it. Since oral board scores are calculated in hundredths of points (82.15, 87.63, 90.87, etc), the goal is to keep building on a few hundredths of points here on this question, a few hundredths there on that answer, gaining a few more hundredths with their signature personalized life experience stories at the appropriate time, delivering the all powerful "Nugget" answers that no one else can tell, and pulling away from the parrot salvo -dropping clones.

What's a "Nugget?" A "Nugget" is an answer to an oral board question that will get you your best score (highest possible) on that question, satisfy the board, and cause them to go on to the next question. You will get bonus points by personalizing your answer.

The "Nugget Principle" enables applicants to personalize their presentations to separate them from the "clone" candidates. No one else can tell your story. Once you start lacing answers with your personal life experiences, no matter if they aren't fire department related, is where you start creating the magic.

Before the clone candidates realize what has happened, these candidates have added on extra points to their score placing them in a position to be invited to the chief's interview where they get a real shot at the badge. Just being 1 to 2 points out of the running can decide whether you will go forward in the hiring process or not.

The toughest thing for candidates to do in an oral is to be themselves on purpose. Your stories establish a natural bridge between you and the panel. When you're yourself, you become conversational because you are on your own turf. This alone can lower the stress and the butterflies.

Stories are more than facts. If you can recreate the excitement, emotion, the color and magic to relive the actual event, you will capture the interest and a top score on that question. A big part of getting this job is convincing the oral board that you can do the job before you get it. Stories are convincing and can demonstrate your experience, even if they're not fire related.

One reason stories work effectively is because they go directly to the brain and entertain. They do not require the mental processing of more formal nonfiction writing. Stories have heart and ring true. Collect illustrative stories as you are collecting facts, quotations and other information for your signature stories.

Practice those stories with a hand held voice recorder. Condense them down to a couple of minutes or less. Don't go on a journey. The oral board is not packed for the trip. You won't have time and it's not appropriate to use a signature story for every answer. Tell the story. Make the point. Move on. Once you answer an oral board with a signature story, you can marry the rest of your answer with those clone answers you have been using. Try it and see the amazing difference.

"Put it before them briefly so they will read it, clearly so they will appreciate it, picturesquely so they will remember it and, above all, accurately so they will be guided by its light."—Joseph Pulitzer, (1847-1911) American journalist.

I was coaching a candidate one day and a candidate was giving me those clone answers why he wanted to be a firefighter. I stopped him and had him rewind the videotape of his life to where he first got the spark to be a firefighter. He said, "Oh, I'm from South America. When I was growing up, we lived with my grandfather who was the fire chief of the city. I got to go with him and be exposed to the whole department."

I asked if he had ever told that story in any of his oral board interviews. He said, "No". Why not? I will bet you big money you are a clone candidate right now. But, I bet you also have some personal signature stories that could instantly change your interview scores.

Another Example: I was doing private coaching session with a candidate. He was telling a story about being a federal firefighter in Yellowstone when it burned. The story was not too exciting the way he was telling it. I had to stop and ask, "It sounds like you were trapped?" He was. Now he tells that story and the hairs start standing up on the back of your neck. You're trapped with him. You can smell the smoke and see the embers dropping around you. Does this story make a difference? Please say yes.

Case in point. I talked to a candidate who was dumping only clone answers on the question "Why do you want to be a firefighter?" Then he realized he could begin his answer with a signature story. He remembered a story he could use about a prank being played on him when he did a ride along with his brother. He couldn't believe the difference when he used this personalized signature story at his next oral board.

The story brought smiles and laughter from the panel members. Along with the calls they went on by the end of the day he knew this was the job that blended all his needs. He followed this story with his standard landmark clone answers. This was the first question on his oral. His answer made everyone more comfortable and the interview flowed a lot smoother than before.

"You can't control the wind, but you can adjust your sails."

Can you imagine a Navy Seal Medic who couldn't get hired?

Yep, that's right. This super qualified candidate couldn't get hired. Why? He couldn't pass the oral high enough to get called back.

I coached this Navy Seal. Like many candidates he was stuck in the process. He had a lot of clone garbage. At one point he held up one of those books with all the oral board answers and said, "Should I still use the answers out of this book?" I told him, "Not if you wanted to get hired." On one of his answers I stopped him. OK, now tell me a personal story that can relate to this answer? The story he starts with we throw out. The second story too. No, no, not the clone stuff. Tell me your personalized story that no one else can tell. He regrouped then I saw the light come on.

Mark started telling the story of being in Afghanistan on a Navy Seal Medic mission at night. He took me on the journey with him. The hairs on my arms and neck stood up. It was exciting, emotional. He relived it. I said, oh, my god. That's it. Mark said, you mean I can use that? Please do. It delivers the point in answering this oral board question. It will convince the panel you not only know the answer, you have lived it!

At the end of the coaching session he had a new killer presentation. All his own stuff. Not a clone of everyone else. We didn't give him anything he didn't already have. He was just shown where it was.

I told him if you're not hired on your next interview you will be picked up soon by someone else. His next interview was the following week. I told Mark they're going to be trying to hire him before someone else does. He passed his interview and was called back for the chief's interview the next Tuesday at 10:00 a.m.

On Tuesday at 4:30 pm my phone rings. It's Mark. You were right. I just got the call. What got him hired? Not packing on more credentials. He already had what it took. He just needed the necessary oral board skills.

When were you going to tell Us?

What have you left out? If it doesn't come out of your mouth during the interview, it didn't happen.

You may be leaving important stuff out of your oral board presentations. The following are segments from our son Captain Rob's coaching sessions.

Two recent candidates left out they were Eagle Scouts. Is that important? Yea!

Although this was a firefighter in Kansas this candidate forgot to include in his oral interview that he grew up in the Las Vegas area and was testing to come back home to Vegas.

Military experience can be a big asset if you present it correctly. Most military veterans don't expand enough on their experiences to the panel. Like many this candidate only mentioned that he was in the military. Which branch? Marines. We're you stationed over seas. Yep. Where? Okinawa and Japan. Did they prepare you with cultural diversity classes before you went to Japan? Yes. So you were taught and lived in a cultural diverse country. I guess. A lot better than just I was in the military. When were you going to tell us?

Another candidate mentioned he was in the military. What was your assignment? I was part of the ground crew for Marine One. Isn't that the President of the United States helicopter? Yes. Did you have a security clearance? Yes, because I was around the President. Should I use that? When were you going to tell us?

This candidate only said he worked for a private company that provides fire protection. What do they do? Weapons research, development and testing. Have you been trained to handle emergency situations and suppression with rocket fuels and explosives? Yes. Do you have a security clearance? Yes. What security clearance do you have? I can't tell you. If he couldn't tell us, this is pretty big right? So, when were you going to tell us?

This candidate was asked, weren't you activated for Iraq? Oh, yea. Tell me the story. Well, I was at the firehouse at Columbia Fire College and the phone rings at 11:30 p.m. It was my crew chief from my Air force reserve unit with orders to report at 7:30 a.m. the next morning at Travis Air Force base. We flew out in a C5A transport and I spent the next year in country and did . ... I've now been in 27 countries. Did you learn about cultural diversity? Yes, let me count the ways. When were you going to tell us?

Did you play sports in high school? Yep. I've been playing sports since I was 6 years old. I played three sports in high school. Did you letter? Yes. In all three sports. Were you captain of the team? Yea, baseball. What did you learn? Commitment, being physically fit, working as a team, supervision, recognizing and using the strengths and weaknesses of the team members. Do any of these have any parallels to the fire service? Every one. Have you ever used these in an oral? Nope. Why not? They're golden. Who else can tell the story?

The candidate only mentioned he was a volunteer. After a few questions it was revealed he lived in the fire station while he was going to school and racked up 8,000 on duty hours. Important?

Too many candidates have been told by firefighters to only use EMS and fire stuff in their oral boards. They end up leaving out 30-40% of critical life experiences from volunteering and jobs they worked through out their life, including high school that can demonstrate skills and attributes that could separate them from the other candidates.

There are, however, things you shouldn't say:

Even though you went into the mission field with your church you never know how this might play out with members of the panel when you bring up church or religion. All you have to do is offend just one panel member and it could affect your score enough to be out of the running. Consider just mentioning how you helped people when and where in culturally diverse parts of the world. You better be praying because your competition is.

From Seattle area: I have a few accomplishments that look really good on paper, but it seems like every time I say them in my voice recorder or in practice with my fiancé they sound like I am bragging. Where can I fit these in, or should I at all.

High School Valedictorian

Full Ride academic scholarship to college

Academy - Most Inspirational

Academy - Most Inspirational, Top Recruit, Most Fit

Yada, yada, yada

Like I said I am not telling you to brag, but I do think they say something about my work ethic and willingness to work with, and help others. Reply: The reason they sound like you're bragging is you are bragging.

I know you want to drop those in but what would you think as a panel member hearing these rants of accolades? When I hear a candidate continue to boast like this I think teacher's pet, can I do the erasers, kiss ass, yada, yada, yada, etc. Yea, you can use one of these accolades but don't go to overkill.

When this candidate was asked if he had any questions for the panel he would reply, do you see any reason why I shouldn't get this job, (because a firefighter friend told him to say that)? Everyone becomes an expert when they get hired you know. This did not play well on the other side of the panel.

A candidate in a recent oral told the panel he stayed in shape with arena boxing. Isn't that cage boxing? Yea. This lead to more questions into areas you might not want to go.

Anytime I hear someone is involved in motocross, cage boxing, ice hockey or any other extreme sports I wonder how many times this candidate has had their bell rung or injuries they have already had or will have that could affect future time and sick leave or ability to do the job. I'm not the only person that feels this way. All you need is a little doubt with one rater and it can affect your score to keep you out of the running to be considered.

In response to the tip I got this e-mail:

So would ballet and bowling be a better choice to put on my resume as opposed to ice hockey or other extreme sports?? Would I not want to include sports that are demanding and rigorous on the body as experienced in firefighting? I noticed you left off the bell ringing all American sport of football?? Thanks for the tips.

Reply: A way to present this is to say you are physically fit and can play most any indoor or outdoor sports.

I often heard and read information from candidates and I think and say, "Please don't."

Here are a few recent examples:

A candidate told me they had whittled down the objective on their resume from four paragraphs to one. The one was still way to long. I asked why? Well, I think it's a really good way for them to get to know me better.

Please don't

Understand we really don't have a lot of time to read resumes and seldom look at the objective.

Another candidate asked me if it was a good idea to give out note pads and pens to the panel members when he walked into the room so they could take notes for him.

Uh, please don't

A candidate who wanted to get on as a volunteer because this department hired most of their rookies from the volunteers was turning himself inside out getting to all the stations on all shifts taking advantage of every opportunity for face time.

The volunteer training officer told him the guys thought he was intense but they were going to give him a shot as a volunteer. He then proceeds to set up SCBA and other training with the individual stations. I asked him why? Here again I want them to know about who I really am.

He said his previous job and training was in journalism and the way to get a story is to ask lots of questions. He was starting out asking a lot of questions at his first volunteer trainings classes.

Please don't

You want to remain invisible to those that are watching you. Your actions will speak for you. All ears and no mouth.

I get a call from an agency with volunteers asking if I knew this new volunteer.

Yep, well someone needs to tell this recruit to keep his mouth shut. Coming into stations trying to impress the firefighters. I had a friend call the recruit who then called me. Why are you doing this? Here again, I want them to really get to know me. I've scheduled several ride alongs. How many? Fourteen.

Please don't. Stop now.

A candidate showed our son Rob and me his resume. It was over 35 pages in plastic dividers in a heavy leather binder. It had a resume profile on the inside cover, cover letter, several pages of a resume, certs, pictures of his academy and the wedding and honeymoon photos. We asked him why???? He said I wanted to let the panel really know about me. Rob and I looked at each other and broke out laughing with who cares?

Many candidates think they need to include resume profiles of their strengths or career profiles on their resumes. Here are a couple of examples from resumes sent to me for review:

Strengths: Performs effectively within group atmosphere. Continuously strives for improvement by education and work experience.

OR:

Career Profile: An experienced Firefighter whose professional responsibilities expanded his knowledge in the areas of fire suppression, emergency medical service, rescue, hazardous material, public relations, public education, fire prevention, and decision making. Qualified by:

1. A proven record of success with progressively increasing responsibilities based upon experience, knowledge, and superior work performance.

2. Strong interpersonal skills and a proven ability to work effectively with individuals in all levels.

3. Excellent communication and time-management skills and abilities.

4. The ability to remain calm and utilize deductive reasoning in critical and demanding situations.

5. Committed to promoting a positive public image of the fire service.

yada, yada, yada, blah, blah, blah

The first thing I do when I see these types of statements is line them out and write, "Who Cares?" Understand the board doesn't! Candidates who put these on their resumes tend to be anal. This is the type of resume material that belongs on a regular job interview resume. Not a firefighter resume. We don't have time to look at this mindless garbage and it takes away from the stuff you really want us to read. I'm a one-page resume guy. Don't give me a book.

Key Words and Specific Traits

I was just wondering if the suggested sample questions would have any answers that the panel interviewers are looking for? Specific words? Specific phrases (types of answers)? I have been trying for quite a few years and seem to fall short after the panel interview. My time is running out, as I am now 31 years of age. In previous interviews I have rehearsed answers that have no negative outlook on them. I am assuming that my answers to questions that they ask are not what the interviewers are looking for. I would deeply appreciate your assistance. Daniel

Reply: First, you're not too old. Our oldest candidate was an EMT hired at age 47. Like a lot of candidates it sounds like you're stuck somewhere in the process. Have you ever practiced your answers with a voice recorder to find out how you sound to the panel?

There is a lot of information out there about focusing on specific traits and key words for interviews, trying to target and use those key words to get the boxes checked off on the rating sheets. There are books that teach you the suggested answers, words, etc but they make you into a clone of everyone else. Since these key words and specific traits can change from agency to agency and from test to test with a department how do you know how to target the current key words and specific traits they're looking for?

We've know for a long time that concentrating on trying to get the boxes checked off on the rating sheet for specific traits draws you away from your personal life "Nugget" experiences. Stories only you can tell, and can give our guys the advantage of killing them with their "Nuggets".

If you've ever heard a candidates try to pull off trying to target specific traits it sounds like a dog and pony show because in trying to ring the bells on the so called specific traits and key words with starts, stops and sputters they lose site of who they really are. They end up like you, not the candidate you want to be. They would make a better presentation with their education, experience and personal life experiences signature stories that only you can tell. This is the cornerstone of our program and in the process gets the scoring sheet boxes checked off, building up hundreds of points in several areas, getting a higher score, in a smooth and natural way. It separates them from the other struggling candidates in their attempt to master the rating sheet.

Lateral-Volunteer to Paid

The biggest problem I've seen on oral boards when seasoned veterans take entry level or lateral tests is they can't place themselves in the position they are applying for; that of being a snotty nosed rookie. They try to hammer the oral board with their credentials thinking the board will just hand them the job. Their oral board skills are rusty and antiquated. It's hard for them to remember how it was to be a rookie.

There is a delicate balance here. Leave your time and rank in your locker. You must be humble, place yourself in the rookie position and build a natural bridge to present your education, experience and integrity to the oral board panel. Without this bridge, you're dead meat. This is not easy for many seasoned candidates. An attitude adjustment is needed. Attitude is a small thing that can make the big difference. Remember the position you're applying for.

The seasoned veteran candidate can roar past any of the other candidates if his attitude and game plan are in place.

Volunteer to Paid and Promotional

Do you have to go through all this preparation and auditioning stuff if you're going to an interview as a volunteer going for a paid position, or a promotional interview, and you already know the guys there?

The biggest mistake job interview candidates make in this situation is when they know people on the panel; they don't think they need to do all the work. They figure everybody already knows about them, and they don't have to say anything. Or, entry level, "It's on my resume, it's on my application, and I don't have to say everything." If it doesn't come out of that slot between your nose and your chin (your mouth), you don't get credit. You might as well have never have shown up. This is how important it is to be auditioning for the part. You play your part no matter who's sitting there.

I can't tell you how many times I've talked to volunteers from departments they have desired to work for. They've waited for years trying for that golden opportunity to get hired in that community as a full-paid fire fighter. Then they blow it. Because they went in and saw Paul was on the board. He knows Paul, they bowl together. Randy over there, why he's married to Randy's cousin. My gosh, he knows everything about me. They come out, and somebody else gets their badge! It's devastating. They failed because they didn't present the package. The other candidates did. It's show time, ta dah! You have to bring out the top hat, the cane, step it out, and give the board the complete show. It's you! It's the bright lights. It's Broadway! You gotta make it happen. You gotta make the magic.

When I said this at a firefighters' convention, Dan shared the following:

. . . I went through exactly through what he's talking about at a promotion in my department for the position of lieutenant. I knew all the people on the board including a division chief. I was thinking, "Geez, I've known these guys for 16

years. And, I don't have to say anything, they know me." During my critique afterwards, the division chief said, "You know, Dan, you've got so much going for you, but you didn't blow your own horn. If you would have blown your own horn, you would have said all the things that you got going for you, you'd have had it. Since you didn't say a word, and I can't give you the badge." If you don't say it, you don't get credit for it, period.

Mock Orals

I'm not a big believer in mock orals. I know you or someone you know have friends, some are firefighters and they have offered you mock orals. Or, your fire academy, college program or a paramedic college offers you the chance for a mock oral complete with videotape. Candidates in these programs practice with oral board flash cards. They play the game of you show me yours and I will show you mine answers. This only cements most if not all the candidates into a clone of everyone else. And you thought you were unique?

Captain Rob says **"Too many 'Experts'** will tell you how and what to do to get hired if they were you. That's the problem. **They're not you!"**

You have firefighter friends that have given you mock orals. Your friends can't bring themselves to tell you how bad you are. You know other candidates who have all the answers. If they had all the answers, they would already have the badge in a city that paid well.

I know you don't want to look a gift horse in the mouth, but if you're offered a mock oral ask yourself, "How many oral boards have these guys been on? Have they been on their department's oral boards, ever? Any oral boards outside their department? How long since their last oral board?" I know an instructor who taught the oral board skills at a community college and his department until recently had never given an oral board and to my knowledge he had never sat on one. Go figure.

You can certainly gain exposure and experience of the oral board setting with mock orals. But be cautious on what information is given. Can these friends tell you how bad you really are? Could you tell your friends? Probably not. Add to the experience that you might not be getting the correct information you came for.

Watch Out For the Free Advice

Well it finally happened, after all these years of hearing things firefighter candidates have said in interviews that some expert has told them was the right thing to do, I hear it first hand. I was sitting in the office of the fire station were I was working, the engineer's son had a friend testing for our department and he wanted him to talk to our firefighter, the newest guy on.

I'm sitting there, and from the other room I hear him recommend that this guy tell the board that he wants to be a firefighter because the pay is good and there are lots of days off. Now I'm waiting for them to laugh, and tell him they're kidding. It doesn't happen. The engineer has been on for 26 years, and hasn't had an interview for 19 years. The new guy was a lateral medic, and didn't have much of an entry interview. So I can see how this poor guy can be thinking, he's in a fire station for the department he's testing for, and he's got a guy with many years on, and a guy who was the last one hired. He must be getting the straight scoop. He was getting the exact opposite. He had signed up for the "How To Fail An Oral Board" class, and he didn't know it.

As I walked into the room, I couldn't let this go, the new guy was telling him that a good weakness to share with the board is that you're a perfectionist. Now I've worked around perfectionists and it's no walk in the park, they think they don't do anything right, and neither do you.

The candidate was Hispanic, and I asked him if he spoke Spanish. He told me he spoke a little and could understand a little more. I asked him if that might not be his weakness, that while he spoke some Spanish, it needed improvement. He bought some language tapes on the way home from the station, so he could demonstrate he was doing something to fix the problem.

Now I find myself arguing with the new guy about what the best response is to why you want to be a firefighter. His theory was the board really wants to know why you want to be a firefighter. Trust me on this one, We Don't Care if you like the hours, pay, and status the job will bring you. You need to tailor your responses to match what the board is looking for, not what you feel, save that for your girlfriend. But you can take those things that motivated you to become a firefighter, and make a beautiful response to this question, and then it's your story.

I worked with this same guy, the expert new guy, again the other day. I mentioned to him that I thought his responses were about the worst I'd heard. He said, "Yeah, I've always been lousy at oral interviews." I asked him why he was giving advice and he said, "Well, everyone keeps sending people to me because I'm the new guy, so I figured I'd try to help." I told him he was, if anything hurting their chances, not helping, and he agreed.

Know this. There are people out there who know they are bad, but will still give you advice because you asked.

Types of Questions

In entry-level interviews we are going to ask three types of questions:

1. -Situational questions: to find out how you will handle lying, cheating, stealing, drinking, drug use, and getting along with others.

2. -Information questions: What have you done to prepare. What do you know about our department? These questions have definite answers, it's like a math question two plus two is always four. There is a way, on our ratings sheet, for us to indicate you got it right or you got it wrong.

3. -Subjective questions: Why do you want to be a firefighter, what first got you interested, what is customer service, ethnic diversity, your closing statement. These are questions that do not have a right or wrong answer. We are going to rate you, basically, on if we liked your performance, and if you drew us in. It's more like an English exam; your score is based on you getting us to identify with you.

You want to think of the responses to these questions like a military operation. You want to get in, hit all the targets you can, and then pull right back out. You don't get any points for just talking, and you run the risk of loosing the board.

Take every opportunity you can get to practice your oral interview skills; you can even take police tests. Every time you speak in front of others you will get better and more comfortable doing it. But please understand everybody on the fire department is not an expert, some of them don't even know how they got hired, and after listening to them talk I can't figure it out either. Captain Rob e-mail: captrob@sonic.net

How do I sit in an Oral Board?

Candidates often ask what posture is acceptable in an oral? Is it O.K. to use your hands when in the oral board? Many have taken public speaking classes in high school and college and have learned to use their hands to help emphasis points, etc. They feel pretty comfortable using their hands. Can this help or hurt you in the oral board?

It's O.K. to use your hands. A Stanford University study showed that getting a job is 15% knowledge and 85% percent enthusiasm! How can you be enthusiastic if you're a frozen statue in the chair? It can't be done. If you can light yourself on fire with enthusiasm and bring the excitement, emotional, the magic and color of the actual event in a story, the oral board will stand up, applaud and watch your butt burn.

Candidates often say, they said you can't use your hands. Sit on your hands if you have to. I always ask, "Who are they? Where is that written?" It's perfectly acceptable to use your hands. If you are already using them, you will cause problems trying to stop. You will start concentrating on stopping, which can throw off your timing, concentration and presentation. A good rule of thumb is to keep your hands and arms in the confines of your body.

When I play back a DVD for review during private coaching sessions, many candidates go to a different posture. Some even lean forward placing their elbows on their knees. I look for this different posture. I will ask if that position is more comfortable. If so, I encourage them to use it. It makes for a better personalized delivery in the oral.

I've had candidates in an oral move the chair right up to the table and place their elbows on the table. After a candidate did that in an oral, one of our rater's commented, "Did you see how he took it right to us!"? He got top scores. It was this candidate's style and personality. He was able to pull it off. You might not. It can be risky. If you're going to try it, first ask the raters if you can move the chair.

Also, when you're that close to the interviewers you have to use the owl affect to talk to the raters. That means you have to turn your head way to the extreme right and left to make eye contact and answer the raters questions. This can also violate the raters' healthy boundaries and make them feel uncomfortable.

Sitting in a comfortable boundary for yourself and the raters is critical. A healthy boundary is where others end and you begin.

As each candidate enters the room at an oral board, they have a tendency to pull the chair back a little as they sit down. After several candidates cycle through an interview room the chair can end up further away from the interview table than you might want. If so, ask first, and place the chair where you would feel most comfortable to make your delivery.

Station Visits/Ride Alongs

Ride alongs can help or destroy you! Candidates want the opportunity to do ride alongs as a way of showing interest, gain information for their oral, and can say in their oral they had been to the stations. Often they don't know the culture and etiquette.

We had a candidate in one day for a ride along. He had an opinion on every topic that was brought up, including sports and the current movies. When it came time for lunch, he was the first one on his feet to fill his plate. His mother would have died from embarrassment.

Let me be blunt here. Dummy Up! You don't have enough time or experience to have an opinion! In this situation you have to be humble, have your questions already written down and realize you are a snotty nose rookie. Too many candidates come in wanting the badge so bad they act like they already have time and want to impress the guys with all of their knowledge. BIG ERROR!

This information will spread like wildfire and destroy you with those who will be making the decisions. Too many candidates tank themselves here and they never know what happened. This applies even if you're already a firefighter applying for another department.

Don't take the bait. Even if you have a friend in the station. If the guys want to joke around and play games, don't do it. You are not part of their family yet. You have no time or experience!

Some departments don't allow ride alongs during test time. If you're lucky enough to do a ride along, show up on time with a desert. Home-made is best. Gourmet coffee is always well received. If it's ice cream, make sure it's the round stuff; not the square stuff. We had so much square stuff during one test we had a contest in the back yard to see who could throw the square stuff the furthest.

After giving this information at a college fire program a candidate shows up at my station the next day. He didn't make an appointment, have desert, or have any questions ready. McFly?

One candidate told me in another class that he had made an appointment and had to wait a half hour when he got there. Poor baby. Understand this is our home. We spend more time at the firehouse than with our own family. So here you come waltzing into our home not knowing what to do.

If you're fortunate to get a ride along, stay for lunch if offered. Offer to pay your share and do the dishes. Leave before dinner (unless asked to stay) and never spend the night. You might interfere with the kick back time during and after dinner.

Should you go to as many or all the stations in a department? Please spare us this part. Don't turn yourself inside out trying to cover all of the stations hoping the word will get back that you did. It will make you look anal and compulsive. This will raise its ugly head in the psychological test if you get that far. One or two stations are fine. If you try to do them all, it only increases the chances of saying or doing the wrong thing or catching a shift of malcontents that will badmouth you. If you're bent on doing a ride along, first make an appointment. During test time things get crazy. Be patient. Act like you would if you were the new rookie in the station.

Station Visits: Are they Necessary?

Question: I said that I would always visit a station when I heard a story of a guy who entered an interview and was asked if he had visited any of our stations and did he get the information packet that admin made up? This sounded like a sure fire way to NOT be successful. But after testing a lot it becomes a serious pain in the butt.

I know I will catch a lot of flack for saying that. But think about it, if you take all the tests you can because you really want the job, you could potentially have to visit 3 or 4 different departments in a month. And it is the same routine over time. Call administration for a station numbers, then call the station and explain you would like to visit to ask questions on the dept/training/equipment/etc

And, you must not show up empty handed (unwritten rule). Last time we tested at my department we had people showing up left and right. Some brought stuff, some didn't. It really did not seem to bother anyone. One guy in particular who sat down with us and really gave me a good feeling did not get hired. I doubt the oral board even had a clue that he visited the station. All the information he gathered did not benefit him because I saw the interview questions. So my question is this. Are we just like the mayor running for office and trying to show our faces around and campaign??? What about the buddy system when visiting stations?? Is this looked upon poorly?? Thanks in advance.

Good Points. Yeah, it's not a day at the beach. But, it can be used in part of your answer on what you have done to prepare for the position; that you have been by some of the stations and what you observed; enthusiasm, skills, dedication, willingness to be of service to those trying to become firefighters. You can also learn something unique about the department that few if any of the other candidates can say in their oral i.e., did you know that San Jose has dry-drum hydrants?

Firefighter Academy T-Shirt Theory

As we went into a local restaurant for lunch, here was a guy in a firefighter academy T-shirt. I asked him if he was in the academy or just had the T-shirt. He said he just graduated from the academy. I went to hand him a business card and he said he already received one at a written test. He passed the written and physical and was scheduled for his oral in three weeks. His girl friend then said, "Do you get your hair cut by Ed? Yep." Then I recognized she also worked in the shop where I get my hair cut. As they left, Beth behind the restaurant counter told me that was her daughter. Small world. I asked Beth to have her daughter call me for information on how to improve her boyfriend's chances of getting hired. Many women want to know this because they want a ring and a date to get married and their guy has told them that can't happen until he gets a badge.

A few days later, we were back for lunch at the restaurant. I asked Beth if she gave her daughter the message. I wish I had a dollar for every time I heard this answer: Beth said, "It's covered." I said what does that mean? Beth said, "He's all set. The academy has given him all he needs to pass the oral and get his badge."

I ran this story by Tony who I was helping on his first promotional test. He said that's the same way he had felt, what took him so long to get our program and do the private coaching when he was trying to get hired for entry level a few years ago.

Tony said candidates coming out of the academies are pumped up like Marines with their bulletproof academy T-shirts. They have been given mock orals, use 3X5 flash cards with oral board questions and answers, run these by their friends and think all they have to do is show up for the oral and the job is theirs.

The guy with the T-shirt ended up in band C. This department has never gone to C Band.

Not Yo Momma's Oral

Are you ready?

Are you prepared?

Are all your ducks in a row?

Do you remember where your ducks are?

There has never been a time like this in the fire service. We had 24 people retire from my department in one day this year. We were already short a few. With the enhanced retirement, firefighters can retire as young as 50 years old.

The fire service got huge incentives to hire soldiers coming back from Vietnam in 1970-73. Those people now have 30 years on their department, and can retire at over 85% of their salary. The department retirement payments have gone up a lot with the 3 % at age 50, and it's cheaper for the department to run with fewer people and pay overtime.

Eventually they are going to have to replace most of the people retiring. Yes, the gravy train is in the station, do you have your ticket, or are you going to be standing there waving good-bye to the people who scored one point higher on their oral? Don't be the person wishing you'd not gone skiing or fishing when you should have been preparing. Get hired, get through probation, get some real paychecks, and ski in New Zealand, fish in Alaska next year. You owe it to yourself and all the people that have been helping you. Make them proud, and don't miss out. Can you hear that whistle? All Aboard.

So the question is "Are You Ready"? Because nobody is going to do this stuff for you, and trust me the other guys are getting ready. Ask yourself these questions: Have you spent more time in the last few months getting ready for your test or trying to get laid? Have you spent more money on your preparation or on late DVD's? Think about it because the game just got real.

Those guys have 30 years on now and they are ready to go, are you ready to take their spot? Would you be ready tomorrow? Next week? Are you ever going to be ready, because your Momma isn't here to help you and remind you to get ready? You've got to get your rear in gear and spend some time preparing, so you can get the job you always wanted. You've spent enough time just making it to where you are now, you need to get that job that will pay you what you're worth, give you time to do those things you've been planning for years, and start getting the respect you deserve.

So let's show everyone that all the time you've spent, all the nights you slept on the sidewalk for an application, all the classes you took weren't for nothing, but were part of your plan to be happy for the rest of your life. Because if your career in the fire service is anything like mine has been you will be.

Captain Rob web site: http://www.myfireinterview.com   e-mail: mailto:captrob@sonic.net

Think Twice before you Volunteer Information

I'm preparing for an interview. They are hiring three and I am currently third on the list. They are interviewing the top twelve. The only thing I am worried about now, besides the interview, is a misdemeanor charge of disorderly conduct that I received in April of 1997 during a spring break trip.

It occurred at a hotel party and several people were charged. I was asked about this in another interview. Actually they asked if I had anything to tell them about that they could find out if they checked. So I volunteered the information. How much will this count against me? Do you think they will even know about it? I had to do a finger print card at the police station. Thank You, — Dave

Dave: When you start volunteering information the wheels start coming off the wagon. This is something you don't bring up in an interview, even when they ask you, "If you had anything to tell them about that they could find out if they checked." This is something you cover with the background investigation. Your situation is not a big deal. It's how you explain the situation that makes difference. Yes, if you were charged, there is probably a record of it somewhere.

We had a candidate who knew this question was going to be asked in his chief's interview. He wanted to mention that he had not been hired by another department because there were some "questionable" results about continual drug use on a polygraph test. We advised him not to bring it up. It never came up in the background. There was no polygraph test for this department. He was hired.

Knowing a Panel Member

I encountered a touchy occurrence today. I had an oral interview today and was well prepared. Except for one thing, I walked in and sure enough I knew one of the interviewers. We both played it off very well.

This intimidated me more than any other part of any interview I have participated in. I have a question though about this. Do I talk to this person after the testing process and ask how I performed or do I pretend this never happened?

Any comments would be a great help since this is a personal friend. Neither of us knew that we were both participating in this process.

Reply: It is protocol for the person on the panel that knew you to state that he knew you and ask if everyone was comfortable with the situation. If not, excuse himself. If you felt uncomfortable, you could have said you knew the panel member. He probably thought he was helping you. You let that situation psych you out. This situation may have thrown you off. No matter who is on the panel or what you think they are doing, it's show time. Give your greatest performance.

Yes, I would contact that person and ask what you could do to improve your interview skills. In that process, you can find out how you did.

The testing agency usually shows the panel the list of candidates. If they know anyone, they switch them to another panel. I recognized six names on an oral board list on one test. They were moved to other panels. I found out that another seven came through that had our program and I did not know them.

Then I got my reward. I saw the magic happen. Several of these candidates didn't have the best credentials, but they used their personal life experiences (their stories) to demonstrate that they not only knew the answer to the questions, but they had already experienced it. After the candidates left, the other panel members would say, "Did you see how he brought in that experience from UPS to answer that question?" These candidates didn't just get good scores. They got top scores. You couldn't connect one to the other in how they did it, because they were using their own personal life experiences; not a coached clone of someone else.

I had the opportunity of a lifetime watching these candidates shoot past the competition with simple proven skills. If I wasn't totally convinced that our program worked before, I was after watching these candidates blow the doors of their oral boards.

First Impressions

How can I make a really great first impression to carry me through the interview?

After seeing clone after clone after clone candidates, someone will walk into the room and BAM, BAM, BAM they set the room on fire! They nail the answers. When they leave, the raters say, who was that masked man? We want to give them a badge! You can do it too! I haven't worked with a candidate yet who couldn't do it. They just didn't know they could do it. Just minor corrections are usually all that are needed to separate them from the clone pack of candidates.

Hand Shakes. Master the First Impression

I spoke to a group of volunteers who were mostly aspiring firefighters recently. As I was greeting several members before I started, I shook hands with a big strapping lad who had firefighter written all over him. He had that kind of firm handshake, smiles and focused eye contact that can cause an oral board panel to want to hand him a badge.

A few moments later I turned to shake hands with another big guy. His handshake didn't carry the same message. The big problem was he didn't know. No one had told him. I had him go over and shake hands with the first guy. They worked on it for a few minutes and he returned with a more confident handshake.

The following is from Work Your Network, by Joe "Mr. Network" Pelayo:

A UCLA study found that when 2 people meet for the first time they make 20 distinctions about each other in the first 20 seconds, then spend the next 20 minutes finding out whether or not they were right! The same study found that a handshake is worth an hour's conversation between two people, because handshakes are thought to be a judge of your character.

When shaking hands with a female rater don't wait for the high beams to come on in her eyes because of too much pressure. Just match the pressure in her handshake. At the end of the interview they will usually stand and shake hands again. Same eye contact while thanking (by name or rank if you know) them for the opportunity. Use that handshake to make the right first impression.

What is the proper way to address the interview panel?

What order should they be greeted? *(Chief, then Deputy, then H.R. rep.)*

One of the raters will usually come out, escort you into the room. As you stand up run your hands down the back of your suit pants or dress to remove the clammy moisture. When you're introduced to the other panel members don't turn yourself inside out trying to remember their names (they are often on cards in front of them).

Openings

How long do you think you have once you walk into the oral board to hook them into listening to your stuff? Many guess 2, 3 and 6 minutes. You have 32 seconds. In that first 32 seconds of your oral board you come in with what's called the "halo" effect.

In that first 32 seconds the board is checking your appearance (the strongest nonverbal statement you can make is what you wear), choice of words, inflection, voice, eye contact and body language. If you open with a clone answer, you're dead meat. There are six other areas in the oral board where you can recover, but don't count on that happening. Once you see the glaze come over the oral board's eyes, you've lost them and they won't come back. Trust me. Please open using a signature story about yourself.

Candidates have about a 20-minute opportunity for a 25+-year career. The ultimate goal is to have the least amount of distractions in your oral board. Everyone has his or her opinions. It seems once a person gets hired, they quickly forget how hard it really was to nail that badge.

As well meaning as some people are, I don't believe anyone wants to be responsible for a candidate not being able to complete their pursuit for a badge. What might have worked for one candidate doesn't mean it will automatically work for others.

Since oral board scores are calculated in hundredths of points (82.15, 87.63, 90.87, etc), bad or incorrect information can place a candidate less than one point out of the running and put them out of the process. I have seen this all too often.

> *I just had an oral where I was asked to tell them about myself, my training and education. I proceeded to "dump the whole load". Two questions later I was asked "What have you done to prepare yourself for the position of firefighter." I was stumped. I had just told them everything and now had nothing to say without reiterating. My question is: How do I differentiate the two questions especially because I don't know what I will be asked. One of the worst things you want to do is reiterate in the body or closing of your oral board. Sometimes the raters make errors in asking or combining more than one question at a time.*

This might help: The dilemma is, "Shall I have a short or long answer for the typical opening question, 'Tell us a little about yourself'." Remember "a little". This is just an ice breaker question to get you comfortable in the chair. A one-minute or less answer about you and your hobbies is all that is needed here.

They don't need your name (they already have it) and NEVER tell them your age. A "Nugget" here: If they look baffled after your short answer, ask if they want more. They usually won't.

Most candidates make a big error on this question by dumping the whole load on why they want to be a firefighter, what they have done to prepare, why this city and on and on. That's not what this question is about. It's only to get you comfortable in the chair. Then, when the panel starts asking why you want to be a firefighter, what have you done to prepare and the other standard 30 possible oral board questions, you have to reiterate what you have already said. You lose valuable time and points here.

When some candidates start talking in an oral, it's like going on a journey. There may be no final destination. Most panel members aren't packed for the trip. I asked a candidate to tell me a little about himself during private coaching one day. I stopped him 12 minutes later somewhere in Montana. I said you have just used up 12 minutes of a 20-minute oral. What do you think we have time for now?

Don't Ever Say Pay or Benefits!

Most of us, if not all of us, want to become firefighters for some of the same reasons: good money, good benefits, good work schedule, job security, etc. . . .

To those of you that have scored a 95+ on an oral: Have you ever answered pay and benefits to the famous oral question: "Why do you want to become a Firefighter" and actually receive a good oral score using an answer of pay or benefits?

Reply: Anyone who has the nerve to give an answer of pay, benefits or work schedule is committing instant suicide!

It's an unwritten law. We know it on the oral board, but don't tell us. Sitting on a board, I can't believe candidates would still say those things. You're stepping on a land mine here. If anyone has received a good score from that answer, it was a fluke.

It's better to use your own signature story of what sparked interest to be a firefighter. That is an answer they have never heard.

One of our candidates gave that answer about pay and benefits during coaching. He had great credentials; had taken over thirty tests. We suggested he not use that answer and some others. Three days later, he took his next oral. He got his best score and position ever and got hired.

Shot Gun Effect

Is my answer too long?

I just completed my oral interview with a big department. I feel that everything went great. They asked questions that I specifically studied for. Although I do have to say that I was still very nervous. I had practiced with a voice recorder as you had advised and my script came out great.

I do have one question for you though. I was asked the question of, "What do you know about the City of _____ and their fire dept?" I responded with the type of city they are, the location, major freeways, target hazards, square mileage everything imaginable that someone would need to know about the fire department and the city.

As I was giving my answer one of the raters looked up at me and gave a look like he did not even want to hear anything more that I had to say. This concerned me. Should I continue with my answer or somehow try and cut it a little short.

Reply: You give your answer as planned. Although you can't tell what the board is thinking, if the panel looks puzzled ask them if they want more. They will tell if they have heard enough and you can be onto the next question.

The best way to handle this type of question about the city and department is to use the shotgun effect. Give them a smattering of areas like square miles, population, type of city government, number of stations, engines, trucks, number of personnel and target hazards.

What would you think if you were on an oral panel and the candidate gave you a sample smattering answer? Right, you would think he had done their homework.

You don't want to go endlessly here. Just a sample smattering. I had a candidate one day tell us so much he got down to the grid water system the city used. Definitely overkill. Another candidate during coaching had a good answer for city information. In the next two weeks before his oral he piled more information onto his answer. He ended up making a long answer endless, finally telling them the number of convention hotel rooms that were available. He committed suicide in his efforts to over impress the panel. Oh, yeah, this is the guy we want to put in a station that would drive everyone Nuts!

This mindless, endless, rambling not only hurts your score, it robs valuable time that you could be using to let the panel know the important stuff that could improve your chances to make the final cut.

Let's get real here. There are a couple of departments that have an extensive and elaborate protocol that you must follow in order to get hired. You have to literally get married to their process to learn the secret hand shake, the right way to answer, and have to be almost sponsored, related or recommended by a recognized member of the department in order to go forward in the hiring process.

If you followed this intense formula, you might find it doesn't work with the majority of the other departments you will be testing for. I've had some of these jewels on my oral boards. They talked endlessly. They almost put us to sleep. We couldn't wait for them to leave the room. Going on and on only turns off the mind of the individual panel members. This is nothing more than Russian roulette.

We have done detoxification-coaching sessions with some of these brainwashed converts.

This certainly does not go along with my policy of keeping things simple (KISS) for the majority of tests you will encounter. Some postings on firefighter bulletin boards are examples of guys who want the interview to be the way they want it to be and are disregarding the way the interview really is.

The competition became so intense on one of these departments that they attracted a large number of candidates whose career track had been to become or were attorneys. Competition is so keen when you take the Bar exam to become an attorney, no matter what the scores are, they drop those at the bottom of the list. This created big problems with the original "attorney wanna be's" once they got into the academy. The typical team-player attitude to help another candidate when they encountered problems was missing. Not only did they not have the team attitude, they would turn on the candidate so they wouldn't be the one dropped. This created additional problems once these beauties went on the floor. I got your back covered was not one of their slogans.

The choice is yours. Absolutely nothing counts 'til you have the badge. Nothing!

Presenting Your Credentials?

How do you present your credentials?

I am a 22-year-old female graduating from college this June with a degree in Exercise and Sports Science.

I have never been to an oral board before, except for my volunteer department (and it was pretty lax). Do I play-up my college education? My fire service experience 2.5 years?

I am currently the Resident Lieutenant; do I put emphasis on that? I am also an instructor for our local training institute. How do I make myself memorable to the board, or can I?? — Jan

Understand that you are applying for a snott nose rookie position. This is a balancing act. If you come on too strong on your education, experience and training you could hurt yourself. This is not a corporate interview.

Too many candidates want to over impress the board. Those that are already firefighters or have experience like you have a tough time remembering this is an entry-level position. You must be humble.

You can include the information in the question "What have you done to prepare for the position", being careful to leave your rank and great status in your locker.

Psyching Yourself Out?

It's all in the four inches between your ears.

About two years ago I worked as a paramedic for AMR for 6 months. While working, I was unable to continue with my college degree (scheduling problems). Consequently, I decided to resign from AMR and work on my degree. I have noticed, however, that during interviews, departments question my move. Is it wise to omit AMR from my applications and resume? I have nothing to hide (I was not fired nor forced to resign). I feel that leaving AMR has, in a way, hurt my oral interview scores. Furthermore, does this problem have a "nugget" answer? How should I handle this?

It's how you present it. If you say you resigned from AMR, it might raise some questions. If you said, "I worked for AMR, and then continued with my education to complete my degree," should not raise any questions.

You could be psyching yourself out when this comes up. It could be that the board wants to clarify the information, not question your move. Everyone knows the crazy schedule AMR works and how it would cause problems with your education.

You can't know what the panel is thinking. Once you start trying, you will tank your oral board score. Just give your best performance no matter what you think the oral board is doing. Trying to interpret the expressions, attitudes of the panel, what they are writing, etc., is mental masturbation. I had several candidates contact me after their orals where I was on their panel. They would tell me what they thought I was thinking or doing. They were never right.

Short Interviews

There are oral boards where they only ask you a few questions. Then they ask if you have anything to add or would you like to give a closing statement. You're missing out here by not taking advantage of a great opportunity to deliver those answers you have not already used, i.e., Why do you want to be a firefighter? What have you done to prepare for the position? Why do you want to work for this department or agency?

This is like going to Broadway to be in a play. You're playing the part of a firefighter. When they ask you if you want to add anything, it's ShowTime! The bright lights are coming on. The curtain is going up for a 25-year career.

A candidate asked this question: I had an interview with a city fire department. I was scheduled for a half an hour interview with the board, they presented themselves and I shook their hands firmly with confidence. They asked me 6 questions. They were worded a little bit different from what I was studying in your program but the same point! I answered them, but I had little stumbles here and there. What worried me was that when I came out of the room there where couple of their city firefighters with stunned faces and I was wondering why? I looked at my watched and saw that I was only in there for about ten to fifteen minutes . . . I'm freaking out here Bob! Are short interviews good or bad? Thanks.

It is not unusual for candidates to stumble a few words in their oral board. The raters expect it. If you had personalized answers for the base line of possible oral questions, practiced faithfully with a voice recorder to hear what the oral board was going to hear out of your mouth, delivered "Nugget" answers, it is not uncommon to get out of an oral board quickly with just 6 questions.

And, don't psych yourself out by thinking their firefighters were stunned. You probably nailed it.

Don't worry until its time to worry. It's not time to worry yet!

B-Pad

In B-Pad, you will be presented with a video that shows many situations. You will be recorded by another video to view how you respond. At the end of each situation, you will be given an opportunity to pick one of several answers or tell what you would do.

B-pad is designed to see how you would react in certain situations if you were a firefighter. If you already had your answers in place, you would get high scores for an oral board, members of a city council hiring committee, B-Pad or any other interview. The problem is most candidates don't have their script down to audition for the job of a firefighter.

We teach candidates to prepare like an actor does for a part in a play. Once they are in the part you only see their personal script on becoming a firefighter. This works especially well in the B-Pad where you would want to act as if you were already a firefighter. Nothing short of this will do. It's show time.

The key in B-pad is to listen and identify the issue, catching more than one issue, deciding the correct thing to do and using the total time you are given to answer each segment.

You can find out more on this process on the Internet by using the key search word: B-Pad or go to bpad.com

Here are four sample scenarios used in B-Pad:

Comforting a trapped child
Aiding an elderly or ill citizen
Responding to conflicting orders
Confronting a coworker's substance abuse

B-Pad candidate testimony:

My advice to B-Pad candidates is to remain calm and relaxed through the process. They will give you 45 seconds to 1 minute to answer. Try to use all the time to answer. If you feel that you answered to the best of your ability and you have extra time, sit there and do not fidget or move around.

Sit still and wait for that scenario to end. Part of the evaluation is watching how you react after you answer. I think I had 8 scenarios that I had to provide answers for. To be honest, they are all common sense. Don't make more out of it than it is. Above all . . . remain CALM and COLLECTED. After all, if you can't do that in front of the camera, how are you going to do it on real calls? Good Luck!!

More on B-Pad here: http://www.bpad.com

Trailing off

I don't want to bore the panel, but I want to make sure that I let them know what I think is important.

That's often the problem. Candidates want to give us so much over-kill information they think is important so there is no question or doubt the panel members understand their answers. The candidates try to give us a blue print, when we just need a sketch. Try to give us a dump truck, when we just need a trailer. It often puts the panel to sleep. Especially if the candidate doesn't know they are using clone answers.

Too many candidates end up rambling towards the end of the oral board answers. I refer to this as trailing off.

Keep it simple. Your mind will tell you when you need to stop. It you don't, you will keep adding unnecessary stuff, mumbling, trailing off and the answer will be pronounced dead way out there somewhere ending with, "Blab, blab, blab. And, that's about all . . ."

When working on an answer or a signature story, use a hand held voice-recorder that goes everywhere your car keys go to practice, rehearse, time, and condense the story down to where you have a great answer and killer story. Have someone in the know listen to your answers to give you some feed back. Be concise, but brief. Tell the story. Make the point. Move on.

You will know if your answers are too long if the raters start getting a glaze on their eyes or they cut you off (not a good sign). And, yes if you go too long in the mouth you will not take advantage of being scored on all the questions.

There is one question though where you will want to dump the whole load, leave nothing out no matter what. That question is What Have You Done to Prepare for the Position.

Shifting Gears on the Oral Board

So, you're going to an oral board next week. Your buddy is taking his oral at the same department three days before you. He's going to give you the questions. You've got it dialed. You're going to nail it! As your oral board starts, something isn't right. The questions aren't the same. The panel has picked up you might know some of the questions and they start switching gears on you. Your presentation is in cement. You start to panic. You're not nailing it. Your timing is off, they catch you flat-footed on some questions, your mouth is dry, your voice goes monotone, and you're forgetting your best stuff. You start to freeze up. You're blowing the chance of a lifetime. What happened?

This could be a double-edged sword. Oral board panels are becoming savvier. We know after the first day some candidates share answers with their friends, even when they sign a statement they will keep the questions confidential until after the test. If we sense you're too quick to answer or pick up your body language that you might already know the answers, the panel will switch gears. Often, we can see the "oh, no" expression on the candidates face when we start switching gears to different questions.

I discourage candidates from sharing questions on oral boards because it causes them to focus on just those questions and set their answers in cement. If you continue to practice the broad base of questions and your signature stories, you can't be caught flat-footed or fooled.

Even when candidates call me when the orals are still going on, I never ever tell other candidates who are going to the same oral the questions. First it's not ethical and my credibility would go right out the window. It could also set them up to have the panel switch gears on them.

Don't Chitty Chat!

A candidate recently asked this question: I recently passed a written and agility for a department and had my oral interview the next day. I felt the interview went pretty well and am waiting to see if I made the final cut.

During the course of the interview I was asked what I have done to prepare myself for this career. I had prepared 4 copies of several training certifications from classes that I have attend and passed them out to the board members. There were 5 certifications in each packet. Was this a giant boo boo? They all looked at the packets and didn't seem to be negative.

I was able to get all of the board members laughing when they asked me if I had any questions and I asked, "How did I do?" This seemed to relax them a lot (all of the board members were firefighters that were at the written and agility test) and we talked for 10 more minutes after the interview was over, very informal with them asking me about the department I currently work for as a paid reserve firefighter, my family, etc. Good? Bad? — Mathew

Reply: Generally handing out material in an entry-level oral is not a good idea because it upsets the normal flow of the interview and takes up too much time. They probably won't read them anyway. The time to have this material read is before you walk into the room. Have it placed with your application before the day of the test.

You don't have enough time to chitty chat. I would be careful chatting after your interview. You might say something that you weren't prepared for (like them asking about the department) or something that could hurt you. You are applying for a snott-nose rookie position. If you came across as too familiar, it could work against you. Just because you work as a paid reserve doesn't give you any time in this situation. In fact it can give you a false impression that you have an advantage.

This was probably true in this case because this candidate did not go forward in the hiring process.

Asking the Panel Questions?

Candidates have been told that you always have to ask a question if you're given the opportunity at the end of an interview or you will lose points. In a regular or corporate interview that might be true. Not true in a fire oral! You never, ever, ever, have a question. We don't expect you to have any questions. I had a guy one day ask, "Since I live so far away, can I start at second step pay to help pay for my gas?" If that question is asked (here's the "Nugget") you can pause as if your gathering your thoughts and then say, "No, I think we covered everything."

We had another candidate say, "You have probably heard about the charges against me for stealing over at the college?" No, we haven't, why don't you tell us about it. Here was another candidate who did an outstanding job in his oral and he had to bring this up. His score dropped like a wounded seagull. This is not the time to bring up anything like this. You never bring up a negative item unless the panel does. They probably won't. If they do, have a simple, short (I said simple and short) answer to the situation.

Don't Use Salvo Drop Answers!

This is from my friend and associate Tom Dominguez: Answering the question longer than a two minute response can be considered a salvo drop. A *salvo drop* is where an air tanker drops the whole load of retardant or water on a fire all at once instead of spreading it out. All retardant compartment doors are opened at the same time. This is done when the retardant is needed all at once. Do you need to "salvo drop" the interview panel on every interview question? One exception is the answer to the question, "What Have you Done to Prepare for the Position?" You don't want to hold anything back here. Dump the whole load.

Tom is right about salvo drops. I've had candidates where the instant we would finish a question they would immediately start like a parrot on this salvo drop, never coming up for air or giving the raters an opportunity to interact. Often it was word for word, without being personalized to the candidate, out of one of the many books out there with suggested oral board answers. Valuable points are lost here.

Keep in mind too that in a 20-minute interview you will have about 5-6 questions and answers. One candidate who arrived late at one of our seminars leaped at the opportunity to answer the next oral board question. Once he was in the hot seat he was asked the question, what do you know about our department?"

The candidate proceeded to give this fast, rapid fire, never coming up or air, long endless salvo drop answer. It was like he was trying to cram everything in he could think of down to fine details. Just when you though he was coming in for a landing, he touched down and took flight again. You could see the glaze coming over those in the room (as you would see from an oral board panel) as he continued.

When he finally ended the first comment from the room was, "Wasn't that answer too long?" The attendees saw first hand how these long endless salvo drop answers can start to work against you to the point of overkill, making you sound anal. Oh, yea again this is the guy we want to stick in a station and drive everyone else crazy.

One candidate said he had been told by many other candidates and firefighters to keep answering until they stop you. Well, put yourself in the position of a panel member and you have to stop this guy to get him to shut up. How would you rate him? If you go endless in your answers, you might get cut off before you got to deliver some of the best stuff.

This goes along with what a BC with a large agency who hadn't been on entry-level interviews for a while said, "In one word "Boring", with few stellar characters. We need to set a time limit. Even though there was a 20-minute allotted time, HR would not let the panels cut any one that wanted to go longer. How can a 22-year old talk for 40 minutes on endless stuff they have never seen and done? Make the point and end your answer please. It's day 12 and we're tired."

Once again, since oral board scores are calculated in hundredths of points (82.15, 87.63, 90.87, etc), the goal is to keep building on a few hundredths of points here, a few there, pulling away from the parrot salvo dropping clones.

Instead of a salvo drop, you're trying to get the panel members to banter back and forth with you on these situational questions. This allows you to satisfy the panel members, get the top score possible on that question, and cause the raters to go on to the next question. You will get bonus points if you personalize your answers by delivering a nugget answer story relating how you have already experienced this situation; even if it isn't fire department related. Creating the banter back and forth gets the raters involved and a chance to deliver a nugget answer. This is one way that candidates are blowing past the competition in the oral boards.

Here's a sample question: You have been given the responsibility of conducting a fire prevention inspection at a business. The business owner is adamant about not letting you enter the business. What would you do and why?

Here is a shortened salvo drop answer: I would start off by asking if this was a bad time. I would attempt to schedule a time with the owner that might be more convenient. If the owner continues to be adamant about not allowing the inspection I would explain the inspection is to ensure the safety of his business, employees, and customers and is required by law. I would also advise him that compliance to the fire code may protect him or his business from liability if a fire were to happen. If the business owner continues to refuse the inspection, I would advise that I would be required to report the situation to my Captain, and advise that I anticipate the department would follow the policies and procedures for non-compliance. I would further explain the specifics of these policies or procedures in hopes the business owner may change his mind and schedule an inspection. All the while I would be friendly and empathetic to the inconvenience this may cause the owner.

Let's take this salvo drop apart and get the banter started with the raters.

Point: You can give your answer to the rater who asked the question as if they were the business owner to try and create the banter.

Candidate: I would start off with asking you if this was a bad time. I would attempt to schedule a time with the owner that might be more convenient.

Rater: -What if the owner is adamant about not allowing the inspection?

Candidate: If the owner continues to be adamant about not allowing the inspection I would explain the inspection is to ensure the safety of his business, employees, and customers and is required by law.

Rater: -What if he still refuses?

Candidate: I would also advise that compliance to the fire code may protect him or his business from liability if a fire were to happen.

Rater: -What if he doesn't care about this stuff?

Candidate: If the business owner continues to refuse the inspection, I would advise him that I am required to report the situation to my Captain, and advise I anticipate the department would follow the policies and procedures for non-compliance.

Rater: The owner still doesn't get the message?

Candidate: I would further explain the specifics of these policies or procedures in hopes the business owner may change his mind and schedule an inspection. All the while I would be friendly and empathetic to the inconvenience this may cause the owner.

Candidate Bonus Points: I experienced this situation while working at XYZ company and I resolved it this way . . . blah, blah, blah.

See the difference? Which do you like best? This will vary from question to question. If the raters don't pick up the baton and start the banter, you can play both parts, i.e. Candidate: I would ask the owner if this were a good time. Candidate playing the owner: If he was adamant about allowing the inspection, I would explain . . . and on and on until you come to the conclusion.

If you are doing banter with the panel members and they attempt to go onto the next question and you're not completed with your answer, ask them if you could finish your answer first. Practice this skill with your hand-held voice recorder.

Now that you're armed with this information, is this going to make a difference in your next oral? Think about it. The salvo dropping clone candidates (yeah, like you used to be) are parading through. Then, you walk in the room and it's show time. You're creating banter while getting top scores on your answers. Unknowingly, you are causing the raters to say in their minds, that's it! This is what we have been waiting for all day.

More Difficult Questions

Cultural Diversity

We often hear, "I'm stumped on cultural diversity questions."

For most candidates, the oral board is the most difficult part of the hiring process. Cultural diversity questions can be the toughest. Most candidates have trouble in this area. Even those with diverse backgrounds. It's not as easy for most people to just use common sense. We don't give answers to questions. If we did that, we would make candidates into clones of everyone else. Yes, we get the wheels turning in the candidates mind.

Many don't realize they have already experienced cultural diversity going to school, in the military and at work. Once they can identify situations that demonstrate that understanding, and put it in a story form as part of their answer to this question in an oral board, is where it starts making a difference.

I don't want to encourage anything that would make you or the other candidates into Clones. This is an area where you have to have a personalized answer. Cultural diversity is covered in our private coaching sessions.

We use a simple three-step method:

1. Giving a definition of what cultural diversity is to you

2. How we have to be as firefighters in a diverse area

3. How it will be to live in a diverse firehouse.

Then lacing these areas with personalized stories how you have already lived in diverse situations.

Testimony:

I was having trouble with the cultural diversity questions. I scored a 78 in Denver after a poor answer in this area. I asked Captain Bob if he knew anything about cultural diversity after a class he presented at our college. He did! I learned I had a personal dynamite answer for cultural diversity and the rest of my oral board skills. Three months later, I went to the chief's oral in Las Vegas. I was ready when they asked the cultural diversity question. After the interview, one of the Chief's said, "I don't know if we're supposed to say anything, but that was the best answer we have heard on cultural diversity."

Thirty-five hundred took the test. My oral board score jumped from 78 to a 96! I was one of the first hired for next academy. Since then, they sent me to medic school. Believe me when I tell you I would not be in uniform right now if it wasn't for "Captain Bob". Sincerely, — Rafael

Scenario Questions

Do you think you have what it takes to answer all situation questions correctly? . . . answer this (in less that an hour)?

What would you do as a rookie firefighter? Your captain asks you to come in his office to review your final evaluation of probation. You notice a smell of alcohol on his breath. How would you reply?

This is a perfect example how you can be fooled on a scenario question. I believe there are only 30 oral board questions. They can be disguised in hundreds of different ways. This is one of the disguises for drinking on the job, which is number 12 on our 30 plus list.

Here is a simple way to break a disguised question down. Dissect the question down to its simplest term, one word, of what the question is really about (i.e. stealing, drugs, drinking, etc.). Once you have removed the disguise, you can place it in one of the 30 plus oral board questions you already have answers for.

This is one of the simple tools we have to uncomplicate the oral board process.

One way to help you do this is picture a piece of paper in your mind with a line drawn down the center. On the left of the line are issues dealing with ethics, such as stealing, drugs, or drinking. With ethical issues, you ask appropriate questions to determine what you suspect.

If true, you don't deviate . . . you go straight up to a supervisor. On the right side of the line is anything to do with getting along with others; you will go to great lengths to work it out before going to a supervisor. If you can decide what side of the line the question belongs, you have a better chance of knowing how to answer the question.

So take off the disguise that this is your captain. Dissect the question down to its simplest form; one word. What is this about? Right, drinking. What side of the line is this on? Right or left. If it's on the left side of the line what do we do? Drinking is not tolerated. Right again. Ask questions to determine if your suspicions are correct (are you drinking?). If so, you go straight up (why don't we go to our supervisor) no matter who or what rank is on the other side of the table; and stick to your answer no matter what. YOU WILL NEVER BE WRONG! TRUST ME!

Here's another way this question can be disguised:

You go in the locker room and see a fellow firefighter drinking something that looks like alcohol. What do you do? The clone, soap opera answer would be: I would try to get him into the day room, play cards and try to smell his breath; or I would have him go home sick, or have another firefighter come into relieve him.

These are all soap opera answers. Unfortunately they are taught in fire academies, books with suggested answers and fire technology programs. They will make you a Clone candidate. Don't go on this journey. They are insulting to the oral board. You will loose valuable points here. We are intelligent beings on the other side of the table. Give us credit for that. Don't start a soap opera. Ask a question that would verify your suspicions and give a direct answer; not a soap opera.

Understand that if the oral board fires up a question that sounds like drinking on the job, it's going to be about drinking on the job. If it's a question that sounds like taking drugs on the job, it's going to be about taking drugs on the job; It's not going to be aspirin. If the question sounds like it's about stealing on the job, it's going to be about stealing on the job. If they fire up a question that sounds like sexual harassment, that's what it's going to be about or they wouldn't bring it up.

If they fire-up these questions, take off the disguise ask questions to verify what you suspect, decide what side of the line it belongs on and then take action in fantasyland. Don't be like so many candidates by starting a soap opera.

In reference to the above, several candidates have called me about the following question: You're a firefighter and the guys in the station want to you to participate with them to play a joke on the female firefighter. What do you do?

Many candidates have said, "Well these kinds of jokes are part of being a firefighter, or we are a family; it's expected, or depending how bad the joke is or I've played jokes on others before, etc."

Let's go back to the original formula. Dissect the question down to its simplest form to one or two words. What is the core purpose of this question? Take off the disguise and you will have one of the questions from the 30 plus oral board sample list.

Have you got it?

It's . . . . SEXUAL HARASSMENT!

Many of the candidates screamed out loud when they finally figured out the real purpose of the question. Too many gave a poor answer.

In reference to the above sexual harassment question I received this question:

> *I'm not trying to be cheeky, this is a serious question. Would the board ever ask the question in reverse? Meaning would they ask a female candidate if they would participate in a joke on a male candidate? If yes, would they be looking for the candidate to reply no, that it would be sexual harassment? Would it be appropriate for either gender to say they would participate, distinguishing that only if the joke would not be of a nature that may be sexual harassment? This issue is very important in my city, because they are on the verge of hiring the first female firefighter. Not only do male candidates want to make it understood that they are able to work with females in this environment, likewise I would not want to appear as if I was concerned about this type of scenario or uncomfortable and oversensitive about the cajoling that goes on.*

Reply: Sexual harassment questions are the most dangerous of oral board areas. It's a can of worms. You need to keep your answers short and simple here. Otherwise, you will tank yourself big time. It has nothing to do with which gender, it's how it's interpreted and received by the person that is offended. Now what's your answer?

So in this fantasyland environment, sexual harassment is not tolerated. If you try to draw a line at cajoling, where does it stop? At sarcastic comments, physical contact? You will open a can of worms trying to make everyone happy. You could express how easygoing a person you are, but sexual harassment is not tolerated. Again that is determined by the person who is offended. Male or female. It's the LAW! There are personnel rules to protect you.

Question: Do you see any difference between a male or female supervisor? A simple answer in Fantasyland is "NO". If you say, "I don't see a problem", you're dead. Because, when you use the word problem, it connotes a problem. Fun, isn't it?

One candidate was asked this question in his oral. He replied, "No problem. My current supervisor is a scientist. She is the most intelligent articulate person I've ever met." He had a chance to talk to one of the oral board officers later about his low score. The officer told him when he left the room, the women panel member said, "Did you hear what he said?" The two men went duh? She took it as if all other women weren't as intelligent and articulate as his supervisor. Go figure. This is why you keep it simple here. Now you know why marriage is the only war where you sleep with the enemy.

With this and any other scenario questions you want to confirm what you suspect by talking to the problem person first; always in private. Never jump to a conclusion or assume. If you assume, it will make an ass out of you and me (ass-u-me). If you have a valid problem you are protected.

It's the LAW! There are federal, state and personnel rules to protect you. It is posted on the job announcement and in every firehouse. You will find it listed (you might quote this in your oral) on the job announcement. It reads something like: The City of Cucamonga guarantees its employees to be free from sexual and racial harassment.

If you can locate and quote the personnel section in your oral board, all the better. One of our candidates was asked a harassment question. His reply was, "According to your personnel rules section 2268, sexual harassment will not be tolerated.

Imagine the shocked expression on the oral board panels faces. He had just delivered the "Nugget" answer. They were finished with that question. He must have nailed that question and his oral, because he's wearing that department's badge!

Another scenario: You come into the locker room and see a firefighter going through another firefighter's gym bag. He looks at you startled and puts a candy bar in his pocket. What are you going to do? Many candidates said, "I wouldn't do anything. Hey it's only a candy bar."

Another said I am going to assume the firefighter gave him permission to be in his bag and take the candy bar and it is none of my business and I am done with the question at that point. The "I'm done with the question at that point" kind of just blurted out of my mouth and I couldn't get those words back. Also I wonder if it might have been good to ask the firefighter if he had permission to be in his bag. Should I have asked that?

Oh, yes you should have. What if it was a wallet instead of a candy bar? Would that change your response? Well, yea was the reply. Then treat is like it was a wallet and what is the question about now? Stealing? Yes, it is! It's one of the standard oral board questions disguised in a different way using a candy bar.

Your job is to take off the disguise, ask questions to confirm and then take action. Why? Is stealing tolerated? No. So instead of taking action why start watering down the situation that was given to you (yes, you're being scored by what action you would take)? Well, yea, everyone or "they said" you had to give the person the benefit of the doubt. So you start creating this soap opera of all the possibilities why maybe he could not be, he had permission, or some other make believe soap opera, shuffle off to buffalo, dog and pony show while the panel watches you die a slow death.

Why Do You Want to be a Firefighter?

This is one of the toughest questions to answer without sounding like a Clone.

I have thought long and hard about the answer to "why do you want to be a firefighter". I'm having a tough time putting it into words. My biggest desire to be a firefighter is because I love the way the department functions first as a family, then as a job.

I've had way too many jobs that were just corporate ladder climbers and backstabbers. Do you think if I push the "family" aspect will I get max points for that question?

Reply: This is a "Clone" answer. It will doom your oral board. Try this: There was a point in your pursuit that sparked your interest. It might have been during a class, ride along or a life experience where your mind went click; that's it. This is what I want to do in life. My life is not going to be the same until I get that badge. When did this happen? That's your "Nugget" signature story no one else can tell. Once you have the board hooked into listening to you, you can use those other "Clone" answers to caboose your answer.

By the way, I would never use, "I've had way too many jobs that were just corporate ladder climbers and backstabbers", as part of your answer. It might give a bad impression of you to the oral board.

I don't have any courses or certifications to become a firefighter. Can I still find a personal nugget story that could be relevant and interesting to the oral board without sounding like a clone? H. Barrow

Yes. Use your personal life and job experiences, i.e. customer service, sports, responsibility, working as a team, commitment, challenges, a degree where you learned how to learn, etc. and relate them the job of a firefighter.

I asked a candidate who was testing for Oakland during coaching one day why he wanted to be a firefighter. He gave me the typical "Clone" answer, "It's giving back to the community, public service, helping others, blah, blah, zzzzzzzzzzzz."

I stopped him and asked, "What really got you interested in being a firefighter?" He said, "Oh, well I grew up in Oakland, but moved to Shasta during high school. After graduation I went to hotel management school in Reno. That didn't work out, so I moved back to Oakland and started going to Chabot College. I met an old friend who was in the fire science program. We ended up over at his house. His father was a captain for Oakland. They got me all fired up, I signed up in fire science, got my firefighter 1, became a medic and I'm currently a federal firefighter."

I just sat there amazed. I asked him if he had ever used this (his signature) story before? He said no. You gave me the "Clone" answer and you had this beauty sitting here? He polished up the story and practiced it with a voice recorder. He works proudly for the City of Alameda.

Another candidate remembered he had the Gage and Desoto dish and cup set from the TV series Emergency. His mom had a picture of him in front of the TV as a kid eating off it when the show came on. He took that picture to his orals. Did it work? He works for San Jose Fire.

After a written test I asked a group of six candidates why they wanted to be firefighters. They were amazed that what they thought was unique was only a "Clone". After I worked with one in the group with his signature story of why he wanted to be a firefighter, the rest of the group used the formula to put together their own too.

I have yet to find a candidate who doesn't have signature stories. The problem is they don't know how to use them. You might not know yours today. But, after reading this, you will have some aha's in the next few days.

A Relative in the Fire Service?

My father is a thirty-year veteran of the fire department. My grandfather was a twenty-six year veteran of the police department. Would it be a good idea to incorporate that information into an answer on the oral exam? Thanks Rod

It's tricky. It's a balancing act. It could hurt you. Too many candidates club the oral board over the head with a dad or other relative who is or has been a firefighter. The panel can interpret this as asking for more points.

With my son Rob we used it this way: I've wanted to be a firefighter most of my life because members of my family have been firefighters. He never said who. If they wanted to know they would ask. They only asked once. That department hired him.

We approved the way another candidate used his signature story:

When I was 10 years old, my father as a captain on the Boston Fire Department took me to work with him. That afternoon we got a call. We rolled out with a lot of other rigs to an apartment fire. I saw my dad get off the rig, direct people for rescue and extinguish the fire. I knew right then that I would not be satisfied until I achieved my badge. — Steve

Who else could tell Steve's story? No one. He was there. After we worked on this story in private coaching, Steve was able to recreate the excitement, emotion, enthusiasm, and color of the actual event. You were on the rig with Steve; you saw the flames, and hairs on the back of your neck start standing up. Again, firefighters love firefighter stories. We do. If you can tell the oral board a signature story from your life experience that relates to the answer, it can catapult you past the "Clone" candidates. Steve got hired, then tested and was hired in Boston carrying on a family tradition.

Strengths and Weaknesses

I was going over some questions for interviews, and I was hoping someone could help me with an answer. What are good answers for the question; what are your strengths and weaknesses? What are some bad answers? — *John*

Reply: Let's start with what your answers are first.

O.K. If asked those questions I would probably respond with something like; My strengths are education, willingness to start from the bottom, my diverse background in fields other than fire fighting, and the fact that I have experience but am very adaptable to my current surroundings. My weaknesses are occasional tunnel vision, excitability, and no full-time experience. There are probably a thousand faults, but you get the point. Where do I go from here? John

First understand that if we start giving answers, everyone would clone them and they would lose their value. I encourage candidates do use their own answers, reflecting their personal life experience.

This question can be asked in many ways, i.e.: What attributes do you think a firefighter should possess, or what qualities, what strengths etc. I think you can come up with better strengths. Education, starting at the bottom and a diverse background are not really strengths. They are what you've done to prepare for the position. Areas relating to loyalty, honesty, and being dependable etc. are strengths.

When you're deciding a weakness, use something that might have been a weakness, but you have already done something to correct it i.e., you had a problem speaking in front of groups. You have improved this situation by taking a public speaking class or joining Toastmasters.

Since firefighters are in a living environment, we would not be looking for someone with occasional tunnel vision and excitability. No, full-time experience is not a good choice for a weakness either.

Got a call from a candidate who lives in Washington now and his oral was in 4 days. Joel got his Firefighter 1 from an academy in Southern California. He said it hasn't helped much trying to get a job. He has now been a medic for 8 months with no luck in testing. In the most pathetic monotone voice he said this is the department he really wants to work for and (with absolutely no enthusiasm) he will be one of the 15 hired.

He asked if he could run one of his answers on what a negative is for him that his firefighter buddies and other friends helped him work out. Sure, shoot. Joel said a negative for me is my past. Even though I got a DUI and some other minor stuff, that's not who I really am.

I couldn't believe my ears. Uh, Joel that answer would only open a can of worms. Don't use it.

Joel said, OK how about this one. Another negative for me is my paramedic skills. This job will help me improve them. Again, I couldn't believe my ears. Yep, that's the guy we want to hire, the one with the poor medic skills. Can't use this one either.

As already mentioned, everyone becomes an expert when they get hired. The answers Joel worked out with some firefighters and friends were definitely not helping but hurting him. The bigger problem is he didn't even have a clue. This was just one answer. How bad were the others?

I would like to say this was an isolated incident. But we encounter these bad answers on a regular basis. It is especially painful in an actual oral board where we see the candidates die a slow death one question after another. Then the candidates wonder why they don't get hired. This is an area where we try to keep candidates from stepping on the land mines.

After a little probing, we did find a negative Joel could use that he was working on to improve.

What have you done to prepare for the position?

In your oral boards everything you have done up to that day has prepared you for this opportunity. Too many candidates leave out important life experiences that could make a big difference.

While riding a bike on vacation the chain jumped off both sprockets. Couldn't call the car club and it was a long walk back. I rewound the video tape of my life to when I had a bike and quickly got the chain back on both sprockets, wiped off the grease with a handy wipe and peddled away.

On some departments they will ask tell us a little about yourself and what have you done to prepare for the position. We suggest you still break it down into two questions. One brief ice breaker tell us about yourself and then what have you done to prepare for the position.

Try this: This will probably be your longest answer. Start with your education and keep it in chronological order so you won't forget anything. Then, your life and professional experience in chronological order. Start your experience by rewinding the video of your life to your first and succeeding jobs in life; no matter if you got paid or how menial it seemed. Many have had paper routes, mowed lawns, worked for relatives or at Burger King. O.K., what did you learn? How you learned to work hard, show up on time, have responsibility, provide customer service and how to work as a team.

Many have told me they've been playing sports since they were 6 years old. Did you participate in sports in high school or college? Did you letter? Did your team advance to the regional or state finals? Isn't that working as a team? As a team member you had to stay in shape, have commitment and recognize the strengths and weaknesses of other team members and how you could cover in. Do any of these areas apply to the fire service? You bet! Every one of them. So any time you can relate your personal life experiences in answering an oral board question, you are telling the oral board that you not only know the answer to the question, you have already lived it!

End with those things you can tie your name to. Things where you were part of a team, spearheaded a group, took a project from inception to end or were part of a committee that established a procedure or skill. Include anything you volunteered for no matter when it happened. Once you start on this question you keep going until you finish your answer.

It's critical to practice your answers with a hand-held voice recorder that goes everywhere your car keys go to work it out.

This is how it can play out on a promotional test but also applies to entry-level:

Tony was going for his first Captain's test. During coaching we asked Tony to begin his experience for his answer to what have you done to prepare for this position. Tony's first job was working in a bicycle repair shop. He went through successive jobs and the rest on his experience. At the end of this question Tony told the panel that he spearheaded the establishment (attached his name) of bike paths and trails in the city where he was a firefighter. He also collected, repaired used bikes and gave them to those in need. He also collected donations from businesses to fund this program. This type of presentation is referred to as a recall. Tony came full circle from his first job in life to using the experience years later to establish a community-wide bike program. Tony was promoted to captain his first time out.

It was that early life experience (without the handy wipes) that I recalled to get the chain back on the sprockets and back on the road.

What does Customer Service Mean to you?

Many candidates are being asked this question in their interviews. Most answer the question with giving the citizens great customer service, being courteous, meeting the needs of the community, maintaining the equipment and station, prompt on calls, educating citizens, yada, yada, yada. Most of what they say is part of the job we are expected to do. It's going beyond that creates extraordinary customer service.

So do you have a story that relates to going beyond what is expected on the job? It doesn't have to be fire related. It could be a story from your life experiences. As our son Rob says, "It's something that creates a WOW response when you hear it."

My wife called that her car broke down 50 miles from our house in 110-degree weather. The car just came out of the shop. The shop guy said if you can get it back he would take care of it. If the car went to a dealer in the area, it could cost $1,000. I called our AAA tow insurance. Denise told me that we didn't have the extended tow service on our membership and we would have to pay an additional $7.00 a mile. Ouch. We could add on the extended tow but that wouldn't take effect for 24 hours. Then Denise said, "Wait a minute. You've been a member for a long time. With that she approved the extended towing and my wife and the car were back in an hour." The word WOW fell out of my mouth. That's customer service.

I suggest you get a copy of Chief Alan Brunacini's (Phoenix Fire) book, "Essentials of Fire Department Customer Service". Just being able to say you have read Chief Brunacini's book on customer service is convincing evidence that you are on your game.

A story in the book tells of a crew that responded on a medical call. The guy in need was pouring cement and got chest pains. As the guy was being loaded for transportation to the hospital the crew completed finishing the cement job before it dried. Customer service? You bet.

The doorbell rings at a firehouse. Some guy's trailer hitch ball broke of. One of the guys went out to his truck pulled off the ball and gave it to him. That's customer service.

A crew responded to a diabetic. His house was a mess. The captain first asked if they could clean up the kitchen before they left to make it more livable. With the OK the crew got plastic bags and cleaned up the kitchen. Yes, that's customer service.

An elderly lady who often brought by cookies and was a big supporter of her local firehouse was going out of town. She came by and asked if the crew could feed her cat while she was gone. All three shifts went by to make sure the cat was well fed. Get the picture here?

Fired from Another Department

Let's get real here! One of the most difficult hurdles to get over is being fired by another department. It's easier if you're a medic.

There has been an alarming rate of new firefighters fired in their academy and during probation. The main reasons are new hires are showing up for their academy without being in good enough physical shape. No, just because you passed the physical agility doesn't mean you're physically ready for the academy. You see the physical agility standards over the past few years have been lowered so almost anyone can pass. Fire departments have had little control over this circus. But they know once they get the candidates in the academy they will require them to meet a level of service that the job requires before they put you on the floor.

Others can't remember the manipulative skills for equipment and evolutions. Some seem to want to shoot their mouth off on how they were trained in another academy or department.

Last, too many paramedic firefighters are carrying their position, acting as if they're in charge of patient care no matter what. Understand if you have crewmembers that have been going on EMS calls for over twenty years, they have had to learn something. If your officer tells you something, you had better get quickly focused and ask him or her what they're seeing that you don't. You could be 100% right and they will show you the door.

The main thing to remember is don't bring up getting fired in the oral board unless they do. Too many candidates want to bring it up on their own trying to justify their position and try to do repair work. Don't.

If it is brought up or covered in your background, take responsibility for what you think happened (you may never really find out why), what you learned, and how it has helped you move forward in your career.

One candidate had been let go from an academy. He enrolled in another college academy and made it through with flying colors. When he tested again, he told the background investigator how he improved his skill level and was hired.

Violating a Direct Order

Question: I had an oral a few weeks ago in which I was thrown for a loop. I was asked, "You're a rookie firefighter and your captain tells you to perform a task that you know directly violates the departments policies and procedures, what do you do?"

I answered....SOP's are in place for a reason and they need to be strictly adhered to for safety and also for legal purposes. BUT, as a FF, I am directly under the supervision of a Company officer, and if a captain feels it is pertinent to break SOP to perform an operation, then when I am given an order by my captain, it is my duty to carry out that task, so I would carry out that order.

Reply: This is a dilemma. Some departments want you to blindly follow orders, yes even to the point of jumping off a cliff, and others want you to draw the line at some point where you would remain an asset instead of a liability that would place others in danger that would have to come in after you.

In many situations the panel members aren't from the department you're testing for. So, they might not know the department SOP's. In this scenario type question one panel member is usually asking you this question. If you can create banter back and forth with this panel member, as if they were the captain who is giving you the order, you can start building up valuable points.

The key here is the first part: "you are at a scene and your officer tells you to do something that is a direct order!!!!" Are you going to refuse a direct order from your officer? Our son Captain Rob said, "You could say, I would find it strange that my officer would ask me to do something illegal or immoral so I would I would repeat the order to make sure I understood it correctly." If your officer confirms his order in an emergency situation are you going to take the chance in an oral board and refuse the order?

So, you can start out by repeating the order to make sure you understood it. The captain will confirm that's the order. Then you can banter back and forth as you did above outlining your concern. Once you get to point where he said, (OF COURSE) no it's not a test; he wants the task carried out. Look directly at the panel member and ask, "As my captain are you asking me to violate department policy?" If the answer comes back yes, that's what you will do.

Hey, in real life you might do something else. But the oral board is fantasyland in many ways. Just go through the drill. You could add that you don't know what the Captains plan is or what additional resources they have coming that could be in place before you advance a line, perform a rescue or any other emergency situation.

Closing Statement

The closing and the opening question tell us a little about yourself aren't usually scored. But if you say something good or bad in your closing it could cause the panel to go back to a section that is scored and change it. They could ask you if you have a closing statement or if you have anything to add. Both are the same question.

This is the last time the panel is going to hear from you. There are those who would tell you to raise the flag and beat the drum with a lot of fanfare in your closing statement. Please spare us this part. Understand if you haven't done it in the body of your oral presentation, you're not going to make it up in the closing. REPEATING, IF YOU HAVEN'T DONE IT IN THE BODY OF YOUR ORAL PRESENTATION, YOU'RE NOT GOING TO MAKE IT UP IN THE CLOSING! We had a candidate who tried to show us all his certificates during his closing. McFly?

If a candidate is asked only a few questions or the questions they are asked did not cover the great answers they had for: Why do you want to be a firefighter? What have you done to prepare for the position? Why do you want to work for this department or agency? You're missing out here by not taking advantage of a great opportunity to deliver one or more of those answers in a condensed matter to maximize your presentation to gain a few extra points.

Don't forget that the closing part of an interview is the last time they will hear from you. It's the only time where you can call on the emotions of the interviewers to give you the job. Don't reiterate or try to do repair work on a previous answer. Use only the key points not already covered in your script. Without being boring or lengthy, tell the interviews why you really want the job and, with your qualifications, hope to be considered for the position.

Then shut up and get out of the building. Or, you will say something stupid. We had a guy one day ace his oral. After his closing, he said, "Well, if I don't get this job I can always fall back on that part time-painters job." The panel couldn't believe what this guy just said after acing his oral. Did it hurt his score? Enough to keep him from getting a shot at a badge. Last time I heard, he was still painting.

I asked a class of fire candidates, "What do you want to say if you're given the opportunity to give a closing statement at the end of your oral?" One candidate said, "I would ask them if they saw any reason why I wouldn't get the job." I asked why would you say that. Because that's what you would ask in a regular job interview. Good point. But, understand this is not, repeat, is not a corporate or regular interview. This is a semi-military organization. I told the class I would never, ever ask this question. Hum, do I see any reason why this candidate wouldn't get this job? I do now with that question.

Get Oral Board Coaching!

If you are testing and your results letter doesn't have a score in the hiring range you have to start asking yourself why. Private coaching can make the difference in being over forty on the list and being in the top ten going to the chief's oral to nail a badge!

Here's our formula: Those people who get our program, complete the worksheet, practice with a hand-held voice recorder that goes everywhere your car keys go and come back for a private coaching session with Captain Rob dramatically increase their chances of gaining a badge!!! That's not a statement but a fact!

Did you ever see a batter change his stance after a few words from his coach? And then see his batting average go higher? Of course. As a firefighter candidate coach, I have been there, done that. Got the ball cap, T-shirt and played the game. I've seen beyond the horizon you see.

I'm a coach. A coach is a seasoned veteran, someone who knows the game inside out, upside down, backward and forward. I let these experiences and insights keep you on the high road to a badge.

I have coached candidates who had no experience and those who have been trying up to 18 years to get a firefighter job or promotion. One thing is common with most. After beating their heads against the wall trying everything possible to get this job, they figured out that the real secret is being able to pass the oral high enough to get a shot at the badge.

Can you do it on your own? It's certainly possible if you don't run out of money, jobs, family, friends and hope before you figure it out. Almost immediately candidates who come to us get unstuck, improve their scores and get a real shot at a badge.

Do you spend money on classes? How about books, academies, and training? Then how about the oral board? Oh, yeah I forgot, you're good to go with the oral board right? You've got that wired. You can wing the oral. Right? Then where is your badge? I talked to a candidate who had his chest pushed out saying he was number 15 on this list, twelve on that list and on and on. After several years of testing, he was still another bride's maid.

And, of what you have done so far, what is going to get you the badge?

"There is no free lunch . . . But there are better places to eat."

We show candidates how to bring out their best to make the difference. Sound like fantasyland? Well, why don't you jump on the yellow brick road and add a badge to thousands already hired using our program.

And when you make it big-time and get that badge, as with so many before you, I will bask in the glow of your success. But, as a coach I take no credit, accept no credit. A great coach insists that since you were on the field of play, since you were the one making the moves, the credit all goes to you.

Coaching Plus

The oral board is the most misunderstood and least prepared portion for of testing.

Dennis is an engineer going for Captain. He's sitting in the number 4 position after the written and the first oral. All he had to do was hold his position in the Chief's Oral where there were seven badges waiting. No big deal right? He's stationed with an oral board coach and his captain. His captain is a seasoned veteran rater and trainer for oral boards. They offer to help Dennis. He refuses, saying he's just going to do a lot of reading.

They are also promoting dispatchers. Mary is offered and quickly accepts the offer for coaching. Mary gets the job. On the day of the Chief's Oral for captain, Mary brings in lobster tails and prime rib to celebrate her promotion to senior dispatcher. Dennis thinks he held his position in the oral. The next day people are calling Dennis up to congratulate him on his promotion. Embarrassed, he has to tell them that he didn't get the job. They dropped down to number eight, a firefighter.

Here Dennis had the golden opportunity to gain an edge by taking advantage of coaching from his captain and Dan who were stationed with him.

After many years of coaching, I finally figured out this is a "guy thing" that keeps many guys from being hired. They get great credentials, but do little preparing for the oral.

Coaching must be one of those guy things they can't say. Here's the guys list:

I can't fix it. I'm not lost. I'm sorry. I Love you. And, I don't need coaching.

I know you think the next test will be the one. I've got it now. This adds more stress on you, your family, your relationships, your finances and your credit cards; just a tad more to get that badge. Dream on. I can't take you kicking and screaming to provide you with information. And, I can't take the test for you. The main point I want to make here is TAKE THE POISON EARLY AND GET COACHING.

Glen has been testing about a year. He's 32 years old. Not a medic. He got our program and did the coaching. He had passed two orals and was going back for the Chief's oral for both departments within days of each other. He wanted to come back for another "Coaching Session". Captain Rob did that session. Rob asked Glen what he was going to do if both departments offered him the job. Glen Laughed. The next call from Glen was exactly that, both departments offered him the job. A nice dilemma eh?

When Should I do Coaching?

My question is I would like to do the coaching because I know it would help and I need to improve my oral score. How does the private coaching work? I live in Washington. Do you do it over the phone? If so do you tape the conversation so I can review it or how does that work?

Also, I do not have any orals coming up at the moment so should I do the coaching now and keep on practicing or should I wait until I have an up coming oral to do the coaching?

I ask this because funds are tight and I want to make sure I do it at the appropriate time. Sorry for all the questions, I just don't know how this process works. Thanks for your time and all the great information you continue to supply. — Steve

Reply: We will record your coaching by phone because of time and distance. Too many candidates wait until the last minute to get coaching. You should plan on a coaching session when you have practiced your script with a voice recorder.

We were buying a new cell phone. When one of the sales people noticed my wallet badge, he said, "Hey, Ben is trying to become a firefighter." Ben said, "So is he."

Turns out they both had failed a big city test by just a couple of points. I shared that 60% of that big city test consisted of psych questions (like a lot of entry level written tests now). Had they been to our eatstress.com site, they would have found the inside on these goofy psych questions and passed the written and been part of the candidates left on the list. Big groans.

Ben was taking another oral in 4 days. I asked him if he would like to try out a couple of his answers to oral board questions. Sure. Two quick questions produced garbage clone answers, Ben had been testing for almost two years. I encouraged Ben to check out our web site.

Ben purchased a program I had in my car. He had gone home from work and listened to the entire CD/DVD program. He told me he couldn't believe what he had heard.

The preparation he had done for oral boards consisted of bouncing questions off his friends who were also trying to get on. Since their answers were similar, he felt he had to be in the ballpark. Little did he and his friends know they had become Clone-answer candidates.

Ben called after his next oral. Although he was stumped on the question "what do you know about our agency" (he said that won't happen again), he felt he was 90% better prepared than any previous oral. He said, "Here I was thinking my oral board answers were good enough to get hired and there was nothing more I could do." Ben said going to the web site and listening to the program had been a big eye opener.

Going for All the Marbles

Even though this is for all of the marbles, don't panic now!

I am very excited, and very nervous at the same time. You see I just got the phone call for a chief's oral. Just when I was starting to get familiar with the regular oral interview, it is now time to learn something new! They only gave me a day to prepare. Do you have any pieces of information that might help me? Will the structure be the same?

Should I be studying anything? The city? The organization? IFSTA? Or is this more of a get to know you type of interview? To see how you will fit in. Any advice you might have will help. Thanks for your efforts in helping make people's dreams come true! — Jeff

Many candidates start to panic when they are notified that they are going to a Chief's Oral. They think they have to reinvent themselves. Reinvent the wheel. WHOOAA! Understand what got you there. You are only going to the Chief's Oral because of the great stuff you already used in the first oral. You're riding the winning pony. Don't switch ponies. You're coming around the club house turn, you shoot out from the back of the pack, go to the whip, you're on the winning pony, you're friends and family are on their feet in the stands cheering you on and you ride her home for the badge.

Too many candidates switch ponies because "They Said". If you do not continue to use the good stuff that got you this far, you could drop out of the race. This is a new arena. Candidates who are going to the chief's interview start talking to their friends. They convince them that they need to do something more. By the time of the interview, they're a wreck. It's not them going into the interview. A clone of someone else. The badges are often given to other candidates.

The chief's interview is open to any type of questioning. They are really trying to find out more about you. How you're going to be as a firefighter for the next 25+ years. Do you fit their culture? We like to hire candidates that are themselves on purpose in the interview. Someone who has a personality and is conversational. Are you that person in an interview?

Stan was going to our departments Chief's Oral. He made an appointment to come by our station. While there he asked what more he could do to make it over the top. I told him he was riding the winning pony and not to switch during the home stretch. Three months later I was down at the training center where they were training new recruits. I saw a familiar face. I said, "Stan is that you?" He said, "Yes, I rode that winning pony all the way in!"

Saddle up and ride to glory.

Brass Fever

I screwed up the chief's oral royally last year with drug/alcohol/ and relationship questions. I seem to relate well with some orals that have just firefighters/engineers and a lieutenant/captain on the board. But I definitely have a brass attack when it comes to the chief's. I truly saw my mistakes in your seminar and the CD's/DVD. My scripts are ready this time and the brain computer is running on lots of Gigs...

I'm glad to know you are there in a pinch. I might call on you or your son for some coaching with West Metro. It's not my Broadway but it's really close. I wish I could for Henderson's but it just isn't possible right now. I have sacrificed almost everything except my family's interest and even that gets the stretch sometimes. Sorry not so quick, anything about the service never seems to be. Thanks again. — John

Thank you for your gracious detailed message. As you now know, it's easy to get lost in the process. It's finding the correct formula to get back on the right road that can make the difference.

You can get over white shirt, brass fever! It comes from not feeling prepared, not understanding they really want you to succeed, and being worthy of the position. By practicing with your voice recorder and getting the private coaching will help immensely.

Chief's Oral Errors!

I had passed the first oral board, moved onto the psych test followed the guidelines in the copy of your special report on getting through the psych exam, continued on to the paramedic director interview, which went well until I informed him, when asked, that I did not have experience in Rapid Sequence Intubations. We do not do this in the system where I am a paramedic. I am not sure how well that went over. And on to the chief's interview. — David

You would have not gone forward to the chief's interview if this were a problem.

I rode the winning pony, had the fresh pressed suit, and went in with a winning attitude. The interview went well and seemed to be formatted closer to an oral board more so than a 'get to know you' session. But having been prepared for the orals I had a solid script to work with when presented various questions based on the highlights of my resume. The questions were:

What I had learned about the system and county?

What being a firefighter/medic means to me. Do I have any questions?

Although I was a bit surprised I felt like I had answered the questions with the nuggets phrase even if they rambled on a bit. That's the kind of guy I am, but walked out feeling good that I had made it that far.

The final question in the chief's interview was, "Do I have any questions." My answer was yes! When do we hear about the hiring list or where we, the candidates stand. They just looked at me stoned faced. Think fast rabbit but my real question is due to the progressive nature of the district and how satellite departments are resisting the integration of a fire district vs. city fire departments and blah blah blah- in an effort to show that I was indeed interested in the politics of the area. Could this one slip have ruined it?

This was covered in the CD's/DVD and on our web site. You never have any questions of the panel. Especially bringing up politics! You're applying for a snotty-nose rookie position. You have no time to ask such questions.

Of course as I pick it apart and look for the errors in the final interview, there could be several things that I didn't say that would have increased my score. I thought I would pass them on to you and you might pass them on thru the web site to prospective candidates.

Seldom do you get a chance to include everything.

1) -If there is someone on the board whom was not introduced to you, make sure that you take the incentive to introduce yourself. I can't help but wonder if this was a bit of a test and by not using the persons name on my exit having not been introduced could I have insulted her, lowering my score?

This was not a factor.

2) -I can't help but wonder if wearing the same suit to the Chief's Oral as I had worn to the 1st oral may have cost me some points although it had been a full week in between interviews.

This was not a factor.

3) -If there is a team of liaisons that are present throughout the entire testing process-know their names, positions on the department, and what their presence is all about. They could be grading you on how you interact with the other candidate being a spy more or less. I kept to myself mostly, as advised, did the team encouragement during the physical portion of the tests, but did not do too much bro brah-ing.

In 39 years, other than being on the look out for tattoos, I've never known a department to plant spies to grade a candidate. Keeping to yourself is the best approach.

David didn't receive a job offer.

Medical

Pre-Existing Problems

If you have a pre-existing medical problem or a serious issue in your background, do yourself a favor and find out early if it will interfere with getting a firefighter job.

All to often I receive calls asking "What do I do now?" These candidates have spent time and money gaining education, experience and put their lives on hold trying to get this job, and they have a pre-existing medical problem or traceable background problem that would keep them from gaining a badge.

Situations like not just one but two DUI's. Knee surgery with scars. Back surgery that would show in an X-ray (yes, they are going to X-ray your back) and/or be part of your records with your doctor and insurance company.

Take the poison early. If you have the slightest concern for a medical problem, have the leading expert in this field of medicine (no, not just your family doctor) evaluate your condition. If they feel you're fit for duty, have them give you a letter.

If the situation comes up during your medical, then and only then, produce the letter. The goal is not being DQ'd during the medical and having to fight the doctors later to get back in.

Much has changed since the Aid to Disabled Americans (ADA) went into law. Depending on the situation some agencies will accept a pre-existing medical condition and set a base line for any future injuries. Pretty cool eh?

One of our candidates knew he had a potential problem with his heart that would show up on an EKG. He went to the heart specialist that was a household name in the LA area who gave him a letter that found him fit for duty. Sure enough the EKG revealed the concern, the letter was produced and this candidate works for LA City.

When Captain Rob was taking his medical, the blood work turned up some questionable numbers. We obtained an appointment with the leading blood specialist in the San Francisco area. He determined that the numbers were caused by a recent flu episode. He wrote a letter that cleared up the issue and Rob went on to get his badge.

For other potential problems in your background, have a brief reasonable explanation of the situation.

You don't want to be like a candidate who called about a pre-existing medical problem with his back. They called him Friday for a medical on Tuesday. Monday was a holiday. He had kept his head-in-the-sand in denial when he knew the medical call was the next step. He didn't have many options prior to his medical.

The time to find out is now; before you're going for all the marbles.

What if you don't Pass the Medical?

Pre-employment medical examinations must comply with the Federal Americans with Disabilities Act and in California with the California Fair Employment Act Section 7294.0(d) of title 2 of the California Code of Regulations states:

(2) -Where the results of such medical examination would result in disqualification, an applicant or employee may submit independent medical opinions for consideration before a final determination on disqualification is made.

What this states is that if you have failed a medical or psychological test that was part of a medical, you should be given the chance to obtain a second opinion. Cities and agencies might not be aware of this law. Most people are unaware that they can appeal the decision.

Candidates will say they don't want to pursue this option because it might ruin any chance that this city might call them back in the future or will cause problems applying to other agencies. Although the way the law is written, you can qualify for a list for an agency over and over again; UNDERSTAND that if you are eliminated from the process because of a medical or psychological test, this agency will NEVER consider you again! The time to act is now! And, there are no black lists out there of those candidates who try to get a position that is rightfully theirs.

Eye Surgery

If you're considering eye surgery you should know that more than 5% of patients have problems with night vision, results that don't last and vision worse after than before the surgery.

Here are questions and answers about eye surgery from internet postings:

What are the current attitudes regarding corneal surgery to correct poor vision? My eyes suck, to speak plainly. I'm considering getting laser eye surgery to correct the problem. Do departments consider that "corrected" eyesight? As in, "Candidate must have no worse than 20/70 vision, uncorrected, in either eye."

In general, do departments consider eyes modified by corneal surgery "corrected" or "uncorrected?" Have any trends been noticed?

I'm just looking for how departments are tending to receive candidates who have had the surgery. Thanks!

-Still more: I can't comment on what stand a department may have on surgically corrected vision but I can comment on my own experience. I had Lasik last fall, I was like 20/200 before the procedure. I am now 20/15 in my right eye and 20/18 in my left. It was the best $$$$$ I ever spent. I would recommend it to anyone. Signed Seeing Well.

-Another: I had Lasik in June of last year, and it took like charm. I was 20/100 in one eye and 20/150 in the other. Now I am 20/20 in both, best bucks I spent. No departments have had any problems with my surgery just as long as it was longer than 6 months ago to prove the procedure took. I was never DQ'd (disqualified) in any processes, they just told me to come back for re-evaluation at 6 months past surgery, and I still passed at that time too.

Finally: go for it!

-There were two guys in my academy (paid-new hires) that had their eyes done. The stipulation was they had to be 6 months post op to make sure the correction took, i.e. the lens didn't lose shape after laser surgery. Good luck with your pursuit.

-I had both of my eyes done. Both eyes were -11.5, which I believe is the equivalent of 20/1150. Lasik usually isn't performed on eyes worse than -12 (20/1200).

-You could say I was an extreme case. The day after the surgery I had 20/30 vision in both eyes and subsequently have improved to 20/25 in each eye. I couldn't be happier. I'd certainly recommend it but would encourage individuals to read up on it, research it and definitely check out more than one doctor who performs the surgery. Good Luck!

You can check out more information concerning eye surgery on the Food and Drug Administration web site at http://www.fda.gov/CDRH/LASIK

Colorblind

Colorblind people—color deficient is considered a more accurate term because most of them can see some colors. The genetic defect that causes the visual problem makes it more common in men. Some form of colorblindness affects about one in 12 men and one in 200 women.

One of our candidates was concerned about the color deficient test. Though he had some color deficiency, to his surprise he could identify those colors (they used a multi-colored stuffed animal) that the department was testing for.

ColorMax glasses give patients with color vision problems an option for both subjective and objective tasks they might not otherwise be able to do. But it's highly individualized. Some people do great with it, and others don't get much out of it, and others don't like it. But everybody gets some effect with the glasses, which cost up to $700. Some people go from seeing five colors to seeing 14. By the way, many PD's/FD's are doing away with the CV test all together or are just going with traffic signal color recognition tests. Good luck to you. . . .

The latest NFPA standards section 1582 have an update in the vision section, which states that color blindness will be considered on a case-by-case basis. It was great to see that but also stinks because each department has different standards. You start to find out which departments are more lax than others. Ventura for example is very cool about the issue and has alternate tests such as the Farnsworth 15 which is easier to pass. LA City however is very strict and close-minded to the issue. Only way to beat them is to cheat and I don't want to do that.

From Jeremy:

I just wanted to drop this web site by; it is in regards to color vision. I was recently disqualified from the San Diego fire academy for failing my medical Exam. I failed because I was unable to pass the color vision tests they gave me. I appealed the disqualification and did many hours of research on color deficiency and in my research I came across the ColorMax contact lens called Chromagen lens that are FDA approved. I contacted an ophthalmologist that is provided on their web site and was able to get a prescription to correct my color vision. I was able to pass 100% both color vision tests the Ishihara and Farnsworth. I was able to retest with the city and passed the color vision part of the medical exam. The total cost for the doctors visit and prescription was approximately $500.00. If this can help any one else with color deficiency, I would highly recommend it.

Web Site: http://www.chromagen-international.c...ex2_ingles.htm

Here are some more tips:

Ok, According to the EEOC and ADA if you can't pass the color test given but the doctor asks you to identify basic colors i.e. red, green, and blue etc. You have to show that even though you can't pass the test you will still be able to perform the basic job duties, by being able to tell these basic colors. You just need to articulate to them they you can perform the basic job duties of identifying the colors of cars and clothing descriptions. POST has also said that if an applicant fails the test, but can ID the basics he/she is fine. If the agency will allow it you can get the x chrome contact lenses. The lenses don't enable you to see colors, but changes them so you can identify them.

The test you take is the Ishihara plate test, which tells a Dr. that you are colorblind. The bad thing about this test is unfair for people with a slight colorblindness, since the Ishihara plates are deliberately artificial images, and do not represent the range of colors that people see in everyday life. I had this problem with my agency. Ask them for another option. Ask if you can go and have a Farnsworth D15 test. This test will show how severe your colorblindness is. You will have to pay for it your self. Do keep in mind that some agencies will DQ you for being colorblind no matter what. Good luck and do not let this stop you going for what you want!

Psychological Evaluations

You have to pass the psych test first time out!

Most candidates are more than surprised when I tell them up to 40% fail the psychological test given by many departments.

I received one phone call and two e-mails from relatives of a firefighter/medic candidate who failed a psych test before the candidate called asking "What can I do now?" He had been testing for 5 years and this was the first job offer. I asked him if he knew who we were? Yes. Did you know we had a preparation program for the psych? Yes. Why didn't you get it? I wish I had a dollar for every time I've heard this, "Things were going so great I didn't think I needed it."

Imagine after all the education, experience and time preparing to get this job like the above candidate . . . and you're eliminated. Then no one will talk to you to find out what happened. I've talked to too many candidates who were devastated and didn't know what to do next. This is a critical part of the testing process you need to prepare for and pass the first time out.

You've jumped through all the flaming hoops and made it through the background check. Then, you're conditionally offered the job pending the medical, which includes a psychological test. You take the test, no big deal right? Then the phone stops ringing.

This from a new firefighter:

I want to comment on your psych test information and report. I had to take one for two departments. Well, all I know is that I went into the test and followed your advice. I tried to answer the questions as honestly as I could, while presenting myself as a very positive social person. Some of the "experts" out there say that you should be brutally honest on the test. Well, 3 good guys I know did just that, and they did not pass either test. We lost 10 out of 25 guys on one test! In all honesty I might not have passed either if I hadn't prepared. I feel that is a very dangerous test, and some of the advice these people are giving out is costing great candidates a job. I wanted to let you know that your advice worked, and I owe you much thanks! Steve.

This from an in-service firefighter: During the last hiring process 2 years ago the psychologist passed 10 people. Of those 10, 2 have quit, 2 have been fired, and one committed suicide. I wonder if he is worth what the city pays him to evaluate prospects? Have a nice weekend.

Psychological Interview

The following chapters are segments from our Conquer the Psychological Interview special report. Like every step in the hiring process, if you're headed for a psychological evaluation as conditional offer to be hired you will want to get the expanded complete psychological evaluation report, which includes information on the written portion of the test to be prepared before you show up. The full version of the report is delivered by e-mail. You will be reading the report within minutes of placing your order. More here: http://www.eatstress.com/psych.htm

You have been testing for three, five, or seven years to get a firefighter job, or you are testing for a promotional position in another department. You are finally in the hiring process. You've made it through the background check. Then, you're conditionally offered the job pending the medical, which includes a psychological test. You take the test, no big deal right? Then the phone stops ringing. You are out of the process. You are told that you didn't meet the profile. What profile?

What do you mean, "I didn't meet the profile?" I've got training, experience, education, every degree, certificate, merit badge, and a paramedic certification. I've been a volunteer, paid member of another department for 10 years, and lived and breathed this job. And, I don't meet the profile?

The Personnel Department won't talk to you. They say it was the psychologist who passed judgment. The psychologist won't talk to you. You can't see the results of the test. You're devastated!

The psychological test is changing the face of the fire service. Sure there are some folks who have a lot of baggage and shouldn't be hired. But most of the red-hot's, the backbone of the fire service, can't make it through the process. Surprisingly, the evaluations are based on the performance of those already in the fire service.

More and more agencies are using the psychological test in their hiring process. Psychologists are competing for this lucrative business and agencies feel they need this service to hire the right candidates. In one large department forty-percent of candidates were eliminated from the hiring process through the psychological tests. Fire administrations feel theirs hands are tied and get frustrated when they see that a high percentage of their superior candidates who were eliminated by their psychological test are being hired by other agencies. Some departments have stopped using the psychological testing because they're not happy with the candidates it's producing.

Psych Report

"Psychologists are given more power than they should have," says Robert Thomas Flint, Ph.D., who sometimes does re-evaluations of potential peace officers and firefighters who have failed psychological tests. Although he tends to agree 40-50% of the original decisions were valid, he finds that another 30–50% of the rejected candidates are acceptable and can handle the job.

Dr. Flint feels that the PhD degree has been watered down, i.e., many of graduates in the last ten years, and the psychologist too often paint by the numbers and disqualify a person because they might have an unusual background. These psychologists do not have an adequate background in the statistics and the research necessary to be fully competent in the use of tests with unusual populations. That is, they are trained in identifying problems in the general population but are less skilled in the identifying the strengths in special groups such as firefighters. They also tend to have difficulty incorporating unusual backgrounds into their reports. But, don't a higher percentage of those with a burning desire for this job fall into these categories?

Much of the problem falls on the cities themselves for not having control of the guidelines that the psychologists are using. Left on their own, psychologists will use their own devises to decide what to do, and this is not always related to the department's needs. If the guidelines are not well defined by the agency, then the psychologist might wash the candidate out for reasons not job relevant.

Fire Administrations need to control the process by knowing the target they are trying to hit, what are the traits they are looking for. Then set those guidelines for the psychological test. Some obvious ones are being able to operate in a living environment, conflict resolution, being able to follow instructions, and functioning during emergency calls. Also, if problems of discipline or any other current problems are occurring within the department, those areas of concern should be addressed in the psychological test.

According to Dr. Flint, too much emphasis is placed on the paper and pencil test. He feels strongly that unusual test scores should be evaluated in the light of the candidate's history. Very young candidates 21–25 often do not have enough history to refute problems suggested on the test. All candidates believe of course that they can handle the job, that they can meet any challenges, that they will hold up well at emergencies. The psychologist's job is to determine, as closely as possible whether those beliefs are sound. To give someone the benefit of the doubt maybe endangering them or someone else.

If a candidate can demonstrate that he has overcome areas of conflict that the written test reveals and his early history demonstrates, then the test interpretation should reflect that fact. The paper score then should be thrown out, not the candidate.

A case in point is a candidate that had been a smokejumper paramedic for the forest service. There is no doubt that this person could do the job. He scored high in the initiative category but Dr. Flint wanted to make sure his score on the conformity scale was high enough to balance out his high initiative score. Otherwise he might not be able to follow orders, wanting to do it his way.

This type of situation became apparent when problems were created by ambulance paramedics, who were not aware that they were burning out on ambulance duty and were switching to the fire service. There were serious problems because the seasoned medics became discipline problems when they wanted to do things their way.

The most common written portion of the psychological evaluation is the Minnesota Multiphasic Personality Inventory interview test of up to 1000 questions. The aim here is not to pass the test but to go into the job fully prepared. Put your pride and natural defensiveness aside. They ask a few questions in several different ways. You want to answer questions "strongly for" or "strongly against" instead of being in the middle undecided. Answer questions to present yourself as a more social, interactive, team playing type of person, i.e., you would rather be in a conversation with others than reading a book alone.

The Minnesota Multiphasic Personality Inventory (MMPI2) test compares your way of responding to that of other firefighters or applicants. One of the three creators of the MMPI test felt that it was not the best choice for peace officers or firefighter testing. Many doctors over interpret an indication of emotional instability or they accept a person's average ability as being adequate for the job. Where one category would trigger an earthquake with magnitude of 7 with a doctor in one area, it might really be only a 2.5 on the Richter scale or vice versa.

Before the interview, the psychologist will often have you take a separate personality test, fill out a personal family history, a biography and additional information forms.

Many candidates drive themselves crazy trying to psych the written. You can't. Just take the test as honestly as you can. Remember no one item means anything. Your answers are being compared to successful firefighters. The doctor can tell you how your personality make-up can embrace or undermine your performance as a firefighter. The test is designed to alert the testers if you are trying to psych the test.

You can get some insight on how the written test is scored here:
http://www.bigdeadplace.com/psyche_eval.html

Here is a link to a book with the entire MMPI 2 exam in the appendix and most of the relevant scales.
http://www.amazon.com/gp/product/0816618178/002-9448704-2847250?v=glance&n=283155

Here are the common mistakes candidates are making during the psychological test:

Doctor/Patient Confidentiality

The biggest error candidates make during the psychological test is thinking there is a patient/doctor confidentiality even when the doctor has them sign a release that there is not. This is not your family doctor. Guess who's paying the bill? You should not say anything in confidence during this interview.

Candidates also think that the doctor is looking for people who have never made a mistake or have never done anything wrong. They aren't looking for perfect human beings. That error leads them to present themselves in such an unrealistic positive light that they are seen as excuse makers who will immaturely refuse to accept criticism and correction.

John was going to his first psychological test. His dad was a firefighter. His dad asked him how he would answer the question if he had experimented with alcohol or drugs. John said, "I would tell that him that although I don't now, I did experiment at one time. Hey, he's a doctor and has to hold what I tell him will be in held in confidence." After a couple of heated arguments, John's father convinced him it would be suicide. In fact, if a department refused to hire people who honestly admitted to having experimented with alcohol or marijuana they would have few candidates to choose from. Prior minor use should not be counted against you. But, if you admit that you have, it might.

Extensive use does raise questions. If you admitted to experimenting with marijuana 200 times, this will certainly raise a red flag. Experimenting with or using cocaine, met - amphetamines or mushrooms are much more serious. For some departments these drugs are automatic deal busters. Stating that you never used anything can raise questions. But if you never have experimented with drugs, don't feel that you need to say anything.

If you occasionally drink alcohol and get drunk on your birthday and at a barbeque last summer, that is one thing. But if you're getting hammered a couple nights a week and on weekends, this will raise some questions. Please think first if the doctor asks the question how many drinks would you have at a party and still drive home. Anyone who might answer maybe two could send a message that they will drink and drive. No department wants their firefighters being nailed on drunk driving charges. Responsible people arrange for a designated driver.

Volunteering Information / Don't Create Trails

A doctor who interviews a candidate that is open, honest, and forthcoming, has common sense, and answers all questions probably considers them as O.K. But, many candidates want this job so bad that they will do almost anything to get it. I have been told what candidates have said during their interviews. I've asked, "How did you get these people to say that?" The answer was we just asked them and they volunteered the information. Before you volunteer information, think before you speak. Present your ideas clearly. Don't ramble or chat. Be articulate. This is how you're going to be in the field. Believe it or not this is part of the job interview. You are making an impression of who you are going to be as a firefighter. Make sure you dress up and don't slouch. Be prepared to audition for the part of being a firefighter. Know your strong points. Be prepared to demonstrate you are a team player.

I can't believe what many candidates reveal! Candidates often call me after going to an oral board, doing background checks or psychological interviews. They are concerned by some information they have given. Often it is related to something from their past. My first question is who besides you knows this? Who could they contact that could tell them this information? The usual answer is no one. This is my point.

Why do so many candidates create a trail that could open a can of worms and keep them from getting the job of their dreams? Many feel they have to be honest to a fault to get this job. Candidates tell me, "They were hammering so hard I felt I had to give them something". Please spare me this part. Think twice before creating a trail that no one can find. Especially if it doesn't make any difference.

Don't Be Honest to a fault!

Question from a candidate:

I am about to get hired by a local fire department. The last and final step of this long process is a 5-hour psychological test/interview. I have NEVER taken one of these before. A friend of mine who works with Miami-Dade police says to pass you need to be 100% honest. Any advice or insight is helpful. Thanks a lot.

Those candidates who are honest to a fault diminish their chances of passing the psychological interview! That's right. You folks want this job so bad you will tell the psychologist anything they want to know. Once you start down this road of total honesty, creating trails where you don't have to, you get into big trouble. Especially when the psychologist says, "Everyone has skeletons in their closet, this interview is not designed to eliminate you from the process", or you don't want to be too squeaky clean. So you open up. Then the phone stops ringing and no one will talk to you. You are out of the process Mcfly. And, you don't know why.

A large city fire department called in twelve candidates for the psychological interview. Only three passed. They sent in six more, only two passed. Another six more were tested. Again, only two passed. All those who passed were our people. They got prepared with our program that took us over a year to prepare, to let them know where the land mines were before they went in. Ask them if it was worth knowing the inside secrets?

We have taken candidates who have failed four psych tests, got our report and then passed their next psych interview. This is serious business when you are finally offered a job conditionally on passing the medical; part of the medical is the psych. You can taste that badge. You don't want to blow it when you get this close. Ask the nine candidates above whose dream vanished after their psych interview.

Remember, absolutely nothing counts 'til you have the badge. Nothing! This includes passing the psych!

Even after reading this report, candidates will call me and ask if they should bring up a situation that could not be found. Again, they want to be honest to a fault. Leave no rock unturned. They keep asking and asking again, but should I bring it up. Well, children, it's like this. This special report is just a guide. You can do what you want. But, if you go in volunteering information, creating trails where you do not need to, being honest to a fault, and leaving no rock unturned, don't call me whining when the phone stops ringing and you're out of the process!

One of our candidates was going to a Chief's Oral. He knew one of the questions was going to be, "Is there anything we should know about?" He wanted to say something about being eliminated from hiring by another department because his polygraph was questionable about previous marijuana use. My question was, "What do you think your chances are of being considered by this department it you said that?" Not good was the reply.

My advice was this was not the time to bring something like that up. If anytime, it would be in the background check. Don't create a trail that might not be found. He didn't bring it up. It never came up with the background investigator. A polygraph was not given. I saw him receive his badge. Later, this candidate was hired by the department he really wanted to work for even after taking another polygraph. I witnessed that badge pinning too. The defense rests!

A doctor might say, "You don't want to be too squeaky clean or you will be suspect." Maybe not, but you better take a good look at your record before you answer. Some candidates feel that they have to come up with something. Big, big error. Although there are no halos over fire engines, you had better be squeaky clean.

I knew one of my candidates, Tom, was being considered for the hiring process in a large county department. I told him if he was going to have to take the psychological test to contact me first. He didn't. Tom called devastated with, "What do I do now? I'm out of the process. I asked him if he had thought out his questions before he had volunteered information. He said he had not.

I asked him why he hadn't called me before the test. He said, "Things were going so great I didn't think I needed your help." What do you think now Tom? I told Tom that we don't just concentrate on preparing candidates for the job oral board interview. We are a resource to get candidates to the badge. We shorten the learning curve in every portion of the testing process. We provide the Secret Nugget information for the best book for the written test, physical agility programs, DVD/CD's from our live seminars and coaching by phone on how to "Conquer the Job Interview", background checks, and yes now the all important psychological test. Tom said, "They let me shoot myself in the foot." I said, "No Tom, you shot yourself in the foot." Even golf professionals get lessons to keep their game sharp.

Another department who didn't give a psych test hired Tom. After one year, he had the opportunity to test for the department he had dreamed of working for. He came in for a tune-up coaching session and made the cut for the job, pending the medical that included a psychological evaluation. He panicked. He obtained this "Special Report on Conquering the Psychological Interview". He realized the mistakes he made on his first interview.

As he entered the room for his interview after the written portion of the psych test, he sat in one of five possible chairs directly across from the psychologist. He put on his game face. The doctor had him fill out a personal family history and release forms for other information. He asked him if he would be willing to take a polygraph based on his information. Some psychologists will do this to intimidate you into giving information you do not have to share.

No matter what you sign, understand if it was not listed that a polygraph will be part of the hiring process, they can't demand that you take one. Especially if everyone else is not going to take one. Just go along with the drill.

Tom was honest to a fault in this first psych interview. He knew now not to volunteer any information that would be ammunition to shoot him with. He didn't list or say anything about his crazy family of origin, questions like "any of your parents have a major illness", or toss out anything that could allow the interviewer to get his foot in the door to again take him out.

All the psychologist is trying to determine is the profile that is created by written test. Tom's interview was short this time. This concerned him.

Tom was 36 years old. He called that he passed the psych test and just accepted the job for the city he thought would never come his way. His extended family had rushed over while he was talking to me on the phone. I could hear the champagne corks popping. Tom was crying like a baby. So was I.

It doesn't get much better than this.

Being Open and Honest

When I asked people what they did to pass the interview portion of the psychological test most said they only answered the question that was being asked. They focused on their presentation, instead of tossing something out, not being clear, being cocky, or flooding the interview with partial impressions. They answered the questions appropriately.

Phil was in the hiring process for a big department. At the psychological interview he was asked if he had ever taken drugs. Phil said, "Like a lot of kids in high school he had tried some stuff." He feels that is why he was eliminated from the process. Now Phil says "No". In his mind he is saying I will never do it again. He is twenty-nine years old and hasn't touched the stuff since high school.

Dan was going through his first psychological interview. He was asked about his family. He told the psychologist that his parents probably drank a little too much and his current live-in had been through drug treatment. He was shocked when he found out he was out of the process. No one would tell him why.

Dan was able to obtain a copy of his written psychological evaluation from someone in the system. He was stunned to read that the psychologist felt because of the stresses of a family history of alcohol and drug use and the stressful nature of a firefighter job that Dan would be a candidate for alcohol and drug abuse. This was probably tied with other things relating to the interview. The main point here should have been if there was a pattern in his life. Dan is thirty-three years old now. His life history has never demonstrated this type of behavior before or since that evaluation five years ago. Where do you think Dan would be had he not volunteered the information about his family? The point to remember is to use common sense.

Dan continued to test and four more times he was assigned to take a psychological test with this same firm. You would expect this psychologist to take a fresh look at Dan. Interesting enough, Dan took his fourth test with this firm and the following week, after getting our special report, took another psychological test with another firm. He didn't make it through the original firm again, but passed the new one. I've been told that there is a margin of error of up to 20 percent with these tests. No one seems to want to talk about this. The second psychologist saw a high number in one of the areas of the test that placed Dan in the margin of error category. He took the time to give him another test to verify the error. He passed. He got the badge.

Has Anyone Here Taken a Test Like this Before?

It is not uncommon for the psychologist to ask if have taken a test like this before. If you say yes, you could be asked about the outcome. It is unlikely that a psychologist is going to contact another doctor to obtain your results because no one will rely on a test older than six months. And, a psychologist who would reveal your information to another doctor could be setting himself or herself up for a lawsuit.

What do you think your chances are of making it through a test if you have had problems passing one with another agency? Scott was taking his fifth psychological test. It had been a few years since his first test. He was more mature now and knew the mistakes he had made when he didn't pass his first one. He had been haunted since. Every time he told a new psychologist where he had taken another test he had not made it through. When the question was asked this fifth time though, he remained silent. There is no data bank or secret list that will show where you have taken previous tests. Scott passed this test and now has the job of his dreams.

The Interview

The written test is designed to show a profile of you where possible psychological problems and job difficulties might exist.

There are three general areas:

1. Psychological problems or conflicts.

2. Manifestations of how you express your personality. Are you quiet, a follower, leader, independent, outgoing and friendly without being obtrusive?

3. Assessment. How well you think on your feet. Your ability to maintain a focus when stressed, your judgment, and short-term memory.

If your profile rated you in the severe category in any of the areas, your interview will concentrate on those to verify these findings.

Psychologists can use different rating scales that range from three to seven. They are similar to school grades of A, B, C, or marginal etc. A doctor will generally not disqualify you, but may turn your evaluation back to the administration stating that you are poorly suited for the position. They will let them make the final decision.

As in every field there are people who aren't doing the best job or there are differences in what a department is looking for. Dr. Flint's spin on this is they aren't the bad guys trying to set something up with artificial barriers to do you in. Their first responsibility is to the community and to your co-workers.

His job is to make sure that no one gets on the fire rig that is going to be either a "hot dog" and go where he shouldn't go and get hurt. Or, is going to be careless or panic and someone else gets hurt or dies. The psychologist should be part of a team with administrators, background investigators, and test writers to give input on standards. Then someone has to come up with the final word. All the tests you go through in the process from the written, physical agility, oral boards, medical, background, and psychological are to determine if you are the best person to be on that fire engine.

The evaluation is trying to find variables on a real basic level. Is this guy honest? Does he have integrity? This is partially measured by how straightforward he presents himself. Does he talk like a guy that has nothing to hide? How does he resolve conflict? Does he let things fester into a grudge? What kind of leadership or fellowship characteristics does he demonstrate? Is he disciplined and mature? The main theme is maturity. Unfortunately twenty-two year olds don't have a lot of maturity.

Most of those who make it to the psychological test are solid. A lot of people think that when they don't pass the test that there is something greatly wrong with them. It is not that they have serious shortcomings. All it means is they just aren't scoring high enough to meet the cut on the strengths and high standards they are looking for.

Dr. Flint compares it to being invited to the Olympic camp or being a walk-on at a team practice. The fact that you get that far is a sign that you are pretty darn terrific. But not everybody makes the team.

Dr. Flint rarely sees anyone who has something seriously wrong with them. He evaluated a candidate that stated he wanted the job to get over his shyness. Even though he wasn't a person who locked himself in his room, Dr. Flint felt that he needed to overcome his shyness before he got the job.

Steve took a psychological test. He was turned down for the job. When he tried to find out why, he was told that his profile didn't match the job. When he tried to find out more, he was denied any information. He obtained an attorney and the case ended up in court. He discovered that there was a state law that protected psychologists from having to provide candidates with the results of their tests.

On the stand, the psychologists testified that it could cause psychological problems for the candidate if they read their results. Also, if candidates saw the results of the assessment they would learn how it was structured and be able to spread the word how to pass the testing. Although Steve lost the suit, he was able to receive a copy of his evaluation because of an age discrimination issue.

To Steve's shock, his profile evaluation showed that he was in the severe category for interpersonal difficulty, impulsivity, defensive/evasive, and moderate for antisocial behavior. Nothing in his life history demonstrates these problems. The written summary told a story of someone other than Steve. This could have very well demonstrated the problem of over-interpreting the test data. Even though the psychologist told Steve that he did not pass judgment whether the agency would hire him, right on the evaluation stamped confidential was NOT RECOMMENDED. He lied!

Steve paid to have another psychologist retest him. The second psychologist said he was as straight as a nail. The only question here is if he was comparing him to the average person or the criteria of a firefighter candidate?

Often how a psychologist writes his report is weird. If a person is evaluated that is more likely than the average candidate to have interpersonal difficulties, impulsivity, compulsion, etc., that isn't saying that compared to the general population they are. It isn't saying that they are in extreme risk or displaying these behaviors. It's saying compared to the average firefighter. You don't want someone who is extremely impulsive or antisocial or aggressive to be a real time bomb in the firehouse.

A person can be a little aggressive or abrasive as compared to most people in the population, but that would be a stand out as quite a bit more abrasive than the people he is working within the fire service where guys get along pretty well together and are compatible.

Some people can be a little resistant to authority in the general population, but compared to firefighter candidates he can be bull headed, can't be counted on to do his part in all evolutions of the job or obey all orders not matter what.

Where a candidate with some of these abrasive traits becomes a firefighter, he might express himself in wanting to play his type of loud music in the firehouse. When confronted, he can't understand how defensive his behavior can be. This is not a desirable behavior for firehouse compatibility. The question is how do you look in comparison to the average guy. If you're in the middle of the pack, you're probably going to be O.K.

Tony found out that he had been washed out on the psychological test because he was too confrontational. While explaining the situation to me on the phone, I said, "If you responded in the evaluation like you are talking to me, it is no wonder you were considered confrontational and were washed out." Unfortunately Tony never recognized what he was doing.

Don't Take the Bait!

Because the position of a firefighter is stressful, one of the styles a psychologist will use in the psych interview is to challenge you to see how you react. If this happens, "don't take the bait". Remain cool, think about the question and answer appropriately.

One of our candidates was challenged the moment he arrived for the interview. Everything from his work history, possible gaps between jobs, education, whether he could use the psychologist's pen to fill out a form, and more.

Another candidate was grilled on what he would do if he got lost. The candidate told him he doesn't get lost. When pressed, the candidate told him he got lost once and vowed he would not let that happen again.

Because this job can have challenges on calls and in the firehouse, if the psychologist can get you to get rattled and lose it in the interview there is no way he is going to sign you off for duty.

Now that you know the drill here, don't take the bait!

Psych Challenge

This is a response from one of our candidates that went forward prepared.

I purchased your program for the oral interview and your special report for the psych and was amazed by the whole process. Everything you've said has been right on the money and thanks for your insight.

I just came back from my psych test and I did not "take the bait". He challenged me from my parents to my children and I would not go there.

The following should give you some insight why you need to be prepared for the psych:

The only thing that bothered me was the psychologist tells me first thing is "I would not hire you. You've been in your own business for eight years and still have it, who are you going to be loyal to the department or your business, probably your business." I replied that my wife and I have run my business since day one and I'm confident my wife can handle it." He responds with "For all I know your wife is some air head that doesn't know anything."

I did not give any response to this and he went on to other insults. All the while I remained calm and positive.

Am I correct in thinking that he could not say I'm not suited for this position because I'm a business owner and he personally questions my loyalties? Or is he just trying to get me to respond in some negative way? Sincerely, Danny

You answered correctly; even the comment about your wife. He wanted to see how you would respond under stress like the job. He was trying to see if you were prepared to step away from your business. Good job.

I just recently purchased the report to better prepare me for my upcoming psych test. With your help and assistance I was able to obtain my goal. I GOT MY BADGE!!!!! Thanks again for your time and help. Steven.

Watch Out for the Land mines!!!

Beware of the Questionnaire! Before the interview, the psychologist will often have you fill out a personal family history, personality and additional information forms. He can use this information as ammunition to shoot you with. Ironically, the psychologist doesn't always have your background information, or because of time doesn't look at all the information. He relies on the form you fill out. This is why it is so critical that although you want this job so bad, not to create trails where you don't have to.

Unfortunately, if you have a DUI or other marks that can be found you have to put it down. At that point you want to work on a reasonable explanation of what happened, what changed your behavior and who you are now. Practice this explanation with a voice recorder until you have it condensed.

If you have a DUI, it will be a challenge but not impossible to get a fire job. Many psychologists look at any possible addiction to interfere with the pressures of the job.

After his psych written, Tony received a call that his results were inconclusive. Since he lived in another state, they let him take home the written to complete again. He left a panicked message on my answering machine. Even though our Special Psych Report tells the candidates to not try to figure out what the questions are geared to finding out, Tony not only tried to do this on his first attempt, but was sitting down with his wife no less (right, this will solve the mystery) at midnight trying to figure out what they were looking for in each and every question when I returned his call. Folks this is mental masturbation.

Stop the presses! I told Tony to immediately stop this circus, go to bed, regroup in the morning and retake the test alone not trying to psych each question. After much waiting after he sent it back, Tony got the call he had his badge! Whew!

I've talked to a number of candidates like the following who wished they had a second chance.

The psychologist asked a candidate how often he consumed alcohol. Oh, I have 2 or 3 beers a night after work as I mow the lawn or wash the car. Even though this candidate had read our psych report on drinking and drugs he still said this. When I asked him why, he said the psychologist was so friendly he didn't think it made a difference.

A friend introduced a candidate to me at breakfast that just completed his psych for a large department. His aunt is a psychologist. She gave him the MMP2 written test and told him what they were looking for. I asked, "You don't have to tell me but how did you answer the psychologist when he asked you if you had ever used drugs?" He said, "I told him I used pot around 15 times, which means 150 times to the psychologist."

I asked him how the psychologist could have found that out if he had not told him. Well, he probably couldn't have. If I told you that too often candidates are eliminated from the process because of past drug use would you have told the psychologist? He said, no. I looked at his buddy and asked him if the background investigator came to your house and asked if your buddy ever did drugs would you tell him yes? He said no way!

Another candidate from Oregon was in background for a department that does not give a poly. He's a volunteer for this department. His dad's a firefighter. He wrote on his background packet that he had stolen several items, had sex with a 17-year-old minor when he was age 18, and much more. I asked him how they would find this information out if he hadn't created the trail. He said they wouldn't have.

Not only did the above candidate not make it through background, he was asked to turn in his gear for the volunteers and the DA was considering pressing charges for having sex with a minor. The other two candidates above didn't make it either. All candidates created trails where they didn't have to.

From Tim: The psychologist asked about background information and made me give him exact dates. Things like when was the last time you had 4 drinks and if so how many did you have, when have you had more than 1 drink per hour and did you drive. The Alcohol questions built on each other first he asked:

On average in the past 3 years how often do you drink? I said maybe 2-3 times a month, 2-4 drinks and some months none at all.

Have you ever had more than 1 drink per hour and driven a car? I said yes. He asked when? I said not since I was 19 that's when I got the minor in possession (I learned from that incident and I don't ever take any chances with drinking and driving)

The next question had to do with have you ever had more than 4 drinks in any amount of time? I said yes. He asked something like what was the time in your life that you did that the most I said when I was in college so I would have been 18 and 19 years old, and I would have about 6-8 drinks at times. Then he asked if anyone has ever told you had a drinking problem. I said no.

CB: How would he have known this if you didn't volunteer the information?????????

TM: You're Right! He wouldn't have known. I will keep you informed.

The Psych! The evaluation with the psychologist is another part of the interview process to get hired. You need to be prepared before you walk in. This is why:

Captain Bob, I was invited to a captain's interview, and did well enough to be invited on to the background and psych evaluation. I completed those, but was not invited to progress to the next step - the physical.

When I called to inquire what had happened, they wouldn't give me an answer, but my gut tells me it was the psych because I don't think my background would raise any concerns. Two things that concerned me about the psych: one was the emphasis placed on psychological counseling, the other on arguments. I had recently seen a therapist for two years regarding relationships I've had. The psychologist seemed to really scrutinize that. Am I going to be automatically disqualified because I've seen a shrink?

And I admitted to two confrontations with co-workers in my current job and I feel like that reflected poorly on me. I can't change the past and am naturally a pretty honest person, and I almost feel now like I may not have a chance. Should I keep trying and will the advice in your package help me be able to provide answers to their questions in a genuine and honest, but positive way? Thanks so much. It seemed like you have really helped so many people. Thanks for what you are doing. Best, Adam

CB Reply: If you continue to reveal the information in your psych tests you can expect the same results. If you had read our psych report prior to the interview with the psychologist you would have been better prepared.

"Just because you're paranoid . . . doesn't mean they're not after you."

CB: If you list on your background and psych information that you have seen a psychologist in the past, the psychologist who interviews you is going to want to know why. If your medical insurance paid the bill there is a record. In the process you will sign off your rights and they can look at anything they want.

Tony listed he had had counseling after being married 8 years and two children because he didn't know if he still loved his wife. The psychologist wanted to talk to his personal psychologist. In the process it was discovered that Tony was also in a support group recommended by his psychologist after he got in a heated argument that almost came to blows because of the stress that had been created when job losses were occurring at his job.

Tony was dropped from the process.

Psychologists have asked some candidates why they checked off on the written portion of the psych that they would have more than 4 drinks and drive, had relationships with minors and experimented with drugs more than 25 times, etc. They actually had not answered that way but the psychologists were just checking to see how you would react or get you to change your answer. If this happens think for a moment and say if I did answer the question that way I made an error.

What do you do if you have not passed the psychological test?

Most agencies are not aware that if you are conditionally offered a position if you pass the medical, that a clinical assessment by a psychologist is considered part of the medial examination. Pre-employment medical examinations must comply with the Federal Americans with Disabilities Act and in California with the California Fair Employment Act Section 7294.0(d) of title 2 of the California Code of Regulations states:

(2) -Where the results of such medical examination would result in disqualification, an applicant or employee may submit independent medical opinions for consideration before a final determination on disqualification is made.

What this states is that if you have not passed a psychological test that was part of a medical, you should be given the chance to obtain a second opinion. Cities and agencies might not be aware of this law. Most people are unaware that they can appeal the decision.

There are psychologists like Dr. Flint who do primarily initial evaluations and a few reevaluations of potential firefighters who have failed psychological test and find they are suitable for hiring. Candidates will say they don't want to pursue this option because it might ruin any chance that this city might call them back in the future, or will cause problems applying to other agencies. Although the way to law is written, you can qualify for a list on an agency over and over again; understand that if you are eliminated from the process because of a psychological test this agency will probably never consider you again. But if you take advantage of the law and have a qualified attorney represent you in obtaining a favorable second psychological opinion you could be reinstated. Of course, this is not guaranteed. Do not attempt this without an attorney. This is how this process can work. Dr. Flint just re-evaluated a candidate for an agency that did not pass his first psychological test. Dr. Flint found him suitable for the position.

Just because you have not made it through an evaluation the first time should not exclude you from passing at another time. Dr. Flint has had candidates come back after getting some job and life experiences and maturing. The tests look different and they look and show their experiences.

Here's a personal experience:

I felt that the family life Joyce shared with me during private coaching might cause problems in the psychological test. I did not want to concern her with that at this point. Getting this job is a process you take one step at a time. We deal with the next step once we arrive.

After taking 32 tests, Joyce passed the next three oral boards after coaching. Then the call came for a chief's interview for the big city for which she really wanted to work. Everything was going great. The next step in the process was the psychological test. After the psych the phone stopped ringing. No one would tell this candidate what was going on. I suspected that the psych had eliminated Joyce; I didn't tell her then.

I suggested she be gracious (no one likes to be hammered) and go in to discuss her situation with the personnel manager. Ask them what can be done to improve her chances for this and other tests. She was told that she was a great candidate, but one of the oral board members had some reservations. I think this was a cover because she would not have gone onto the psych had she not passed the Chief's Oral.

Still being gracious and persistent she asked more questions. The questions we told Joyce to ask alerted the personnel manager that she knew the law. Next thing Joyce receives a letter that she is in band B. Joyce called the personnel manager and he revealed that band A consisted of those in the current hiring for the next academy. Band B was the back-up. She was told she was number 2 in band B.

Hold on now because things start picking up speed for this the ride of your life. One of the candidates in the hiring process in band A declined the job. That put Joyce in number one spot of band B. Another candidate dropped out of band A. Joyce was called to go in Thursday for "Another" psychological test. Folks this was the second opinion we were shooting for. Joyce was ready. The same day Joyce completed her medical.

Joyce called that she had just been offered and accepted a position for this large city fire department. Joyce had her badge! She was ecstatic. So was I. You see, this is my reward. Joyce was on her way to get fitted for uniforms. She was to start the academy in two weeks. She told me this was going to be a real Thanksgiving.

All but one (he elected not to pursue the job) of the candidates mentioned in this article are now wearing a badge. The defense rests!

Like every step in the hiring process, if you're headed for a psychological evaluation as conditional offer to be hired you will want to get the expanded complete psychological evaluation report, which includes information on the written portion of the test to be prepared before you show up. The full version of the report is delivered by e-mail. You will be reading the report within minutes of placing your order. More here: http://www.eatstress.com/psych.htm

Polygraphs

The Lie Behind the Lie Detector

The National Academy of Sciences has issued a devastatingly critical report: http://www.nap.edu/books/0309084369/html . It is especially critical of the kind of pre-employment screening used by many fire departments.

AntiPolygraph.org is a website dedicated to the abolishment of the polygraph from the American workplace. These folks were speakers invited by the National Academy of Sciences' polygraph panel to address its members in Washington, DC. They are always happy to discuss polygraph matters with anyone here who is interested.

If you are facing a pre-employment polygraph examination, you need to know how these purported "tests" really work (and don't). It would be worthwhile to take a look at the NAS report referenced above. See also http://www.antipolygraph.org free e-book, http://www.antipolygraph.org/pubs.shtml which may be downloaded as a PDF file or browsed in HTML format.

Are Polygraphs Lying to Us?

Being prepared for every step of the hiring process before you show up will place you in a better position to end up wearing a badge than being caught flat footed wondering what happened when the career you have been intensely pursuing evaporates before you eyes.

If polygraphs are so great why aren't the results admissible in court cases? Criminologists say lie detector tests pass 10 percent of the liars and fail 20 percent of the truth-tellers.

The first time I ever had contact and talked to Scott was 45 minutes after he got the call that he had failed his poly. Needless to say he was devastated. When I asked him what he had done to prepare for his polygraph he said he used the free information from some of the "experts" on some of the firefighter Internet forums. Using those guidelines he said he went in and spilled his guts, just like going to confession.

Just a few minutes into our conversation he realized that he had become too familiar with the evaluator, got chatty, volunteered too much beyond what was requested, was really nervous but thought everything was going just great.

Scott wrote: I chose to take this test knowing I had nothing to hide and being truthful was the best route I could take. The next test I take, I assure you I will be better prepared. I believe Scott is only referring to understanding the process better. I'm a firm believer in preparing for every step in the hiring process before you get there.

According to Doug Williamson a 35-year veteran polygraph evaluator from www.polygraph.com "It is a very serious mistake to believe that you will pass your polygraph or CVSA tests just because you are telling the truth - they are not "lie detectors". Scientific research proves that simple nervousness will cause a truthful person to fail!"

I talked to Scott after he checked out polygraph.com. He realized he had not been as prepared as he could have been before his polygraph and revealed information beyond what was required. If he had it to do over again he would have been better prepare on the process before his evaluation and could have had a better opportunity of passing without compromising his truthfulness.

I echo what Capt. Bob says in that you need to be prepared going in. I went in with the attitude that I was not going to try to deceive them. Well...I was notified that there were some irregularities on a couple of portions. That was after waiting about 4 weeks for the results. Fortunately they let me go back and address the portions in question. Another nerve-wracking wait, and finally word that I passed. Personally I don't think I could go through that again, I'm glad it's over. If I have any words of advice, it would be to research what a polygraph is all about, relax, and do not...DO NOT be anything less than completely truthful. In the end, what I had done was less important to them than that I was forthcoming about it. Brad

DM: I had a poly today. There were six pages of questions asked verbally by him before he hooked up the poly. When hooked up, there where only ten questions asked 3 times in different order each set. All seemed to go well until the end when the examiner asked me, "Anything you want to tell me about the question you had a problem with?"

Me (puzzled) "NO"

Again he asked as he began to take the equipment off me.

Me (really puzzled now)..."I can't think of anything"

Then he proceeds to tell me that there was one particular question that I had elevation on. Should I consider or was this all a smoke screen to get a boring person to confess to something?

CB: Quite common to use this smoke screen to get you to confess to something.

DM: After everything I've purchased (Your program included), I didn't consider the poly. But after that past experience I looked up polys on your site and that is my next purchase. I hope my oversight did not come back to bite me in the end.

Inconclusive: Often candidates are eliminated through the poly with inconclusive results. Not that you failed, but it's the same as you did. Why is that? You didn't fail and you didn't pass? Your results were inconclusive. You still don't go forward in the hiring process. I think the problem again is candidates need to prepare for the poly the same as with any segment of the hiring process.

Randy had the same problem. He took the poly and the evaluator eliminated him with inconclusive results based on his use of pot within the last five years. He swore he had not. Yea, right you say, but that's his story. So, Randy jumps on the Internet and found www.polygraph.com and www.passapolygraph.com He educated himself on what to expect. He took a poly for another agency and passes with flying colors even that inconclusive area about pot and was hired.

Question: I will be taking a polygraph soon (presuming I do well in the interview) and just tonight talked to a friend of mine who was hired last year at the same dept. He told me that they asked him, during the pre-qualification questions if he had received advice on how to cheat polygraphs or had researched it on the Internet. Beware of this. I am now worried about researching anything about polys on the web.

Reply: No worries

From FF/PM1: Yes, I was asked if I had researched polygraphs in my pre-interview, but not in the actual polygraph exam. The examiner asked me why I researched polygraphs? Was I going to manipulate it? I said of course not, but I always study before going into a test, I knew nothing about polygraphs prior to my hiring process and wanted to know what they are about.

While I am a big advocate of honesty in the hiring process, I am also an advocate on RESEARCH prior to entering into a new portion of the hiring process.

This does not make you a "cheater" or "dishonest" for wanting to know what you are getting yourself into, just thorough. I researched every step of my hiring process. I learned everything I could about interviews, medicals, psych exams, polygraphs, backgrounds...not because I was trying to manipulate the system with deception or dishonesty, but I wanted to know what is next. Honesty was the foundation of everything during my hiring process. I used this formula.....If asked; a truthful answer was always given. If asked a question, I would answer the question honestly then be quiet. All other times I sat there with my mouth shut.

I have had friends fail polygraphs, not because they were dishonest, but they were not educated for what they were getting into. They sat down and were honest, but when answering one of the polygrpaher's questions, they would answer the question and then ADD additional info not even asked for. This ultimately led to their demise.

Learning about a polygraph does not mean I am trying to cheat it. Were you trying to cheat by going to fire stations and finding out what types of questions may be asked on the oral interview? No, you just wanted to be prepared.

Be honest, be prepared and be consistent.

Voice Stress Analyzer

The first time I saw a voice analyzer was in a James Bond movie. Bond used mouth spray to throw the machine off. This is not fiction anymore because several agencies have been convinced this is the cats meow. Better than a polygraph. Interesting, the Department of Justice will not use this system. Those agencies that are using them have experienced a higher rejection rate. Candidates who are failing are going on to be hired by other departments who give a polygraph, too. It is thought these agencies use this system because it costs less.

I know several candidates who have elected not take the voice stress test for one agency because of the high rejection rate and that it will place them on some kind of secret black list that would prevent them from being hire by another department. I'm not aware of the existence of any black lists.

One head of HR (who has been educated beyond her intelligence) still hangs onto this tool even with overwhelming evidence questioning the validity of the process. Many experts say the operator of a voice stress analyzer test has the same chance of determining if you are lying as the flip of a coin. Unfortunately, he or she has the same chance of determining if you are telling the truth. Some say get an attorney and don't take the test.

Consider this:

Have you lost a job opportunity with a public service agency because you wrongly failed a voice stress analyzer test? You may have a legal remedy. Most large agencies are governed by state or municipal civil service rules or laws, which make them also subject to the US government's Equal Employment Opportunity Commission rule 29 CFR 1607, the Uniform Guidelines on Employee Selection Procedures. According to the EEOC, all employee selection tools must meet minimum standards, including validation. It is simply a matter of law that departments must use validated tools for hiring, such as the MMPI, CPI, polygraph, urinalysis, intelligence tests or others that have withstood independent scientific investigation. They are specifically prohibited from using unvalidated methods. The voice analyzer technology falls into the unvalidated category. If you took a voice stress to get a public service job, it is a violation of your rights under these EEOC provisions. Contact an attorney for more advice.

How about this:

The first time I was given the test the operator said it was used only to show me how the machine worked. It still counted and I was eliminated. I requested a second opinion. The second operator said I failed a drug use question, which I have never seen, let alone used the illegal substance in my life, nor would I or could I give an explanation for this. So he did a third test. This time I passed the same drug question but now failed a serious crime question of which I had passed on the second one. So I again will not admit to something that I haven't done. I was told by the person that this test is 99% accurate and I argued the point that I got the same questions right then wrong and visa versa. Obviously there is no discussing or changing their views on the matter of accuracy.

A candidate who failed a voice analyzer test was in disbelief when he was told the machine indicated he had sold drugs and had forced women to have sex with him. Neither was true according to him. He requested and was granted a retest with another operator. Same results. Fortunately he is a medic on several other hiring lists.

A Summary of the Testimony before the Texas Legislature Regarding the Reliability and Validity of the Computer Voice Stress Analyzer

A Summary by Victor L. Cestaro, Ph.D.

During my tenure as a researcher at the Department of Defense Polygraph Institute at Fort McClellan, Alabama, between 1993 and 1999, I performed research using the National Institutes for Truth Verification (NITV) Computer Voice Stress Analyzer (CVSA).

To my knowledge, no other scientific research has been conducted using the Computer Voice Stress Analyzer, and there has been no scientific evidence presented that would support the contention that the CVSA is capable of detecting differential levels of stress, or differentiating between truth and deception at any level greater than chance. Unless and until there is compelling scientific evidence to the contrary, it is my opinion that the CVSA is not capable of distinguishing truth from deception in human speech.

Background

It's often not the problem but how you write out an explanation for the background. Most candidates turn themselves inside out when a simple explanation is often all that is needed to resolve most of the situations.

You're in the hiring process. Let's take it one step at a time. The badge is there. There is only one person keeping you from getting it . . . It's YOU!

"Some times I think my mind would kill me if it didn't need me for transportation."

It's not the items of concern, but the when and why. Like most candidates that enter the hiring process, they get hyper-vigilant and the voices of panic and guilt start running around in their heads.

When did these incidents happen? If it was a financial problem, what was your financial situation then?

For other problems and misdemeanors, has the problem happened since?

Was this a financial problem? Many candidates live on a shoestring trying to gain the education and experience to get this job. A couple of 60-days late can be explained if you were between jobs. Are the accounts current now? During the current economic times many are affected. Agencies are going to have to take a different look at credit. Of course the best course of action is to maintain a good credit rating.

One of my candidates had some outstanding bills. In his current background check he told the investigator told how he was resolving the problem. The investigator told him if he cleared them, he could go forward in the process. He did. He went forward in the hiring process.

> I have a background check in the future. I do not have a criminal record nor have I gotten a ticket for the last eight years. I did have a credit card company cancel my account due to late payments. This debt was passed over to a credit collection agency which I paid. I also have some late payments with my health insurance. Living in the Silicon Valley can hit your pockets hard, ya know what I'm sayin!!! Does this hurt me in any way???

Reply: Again, you know the background is coming up at some time. You want to know what your credit report says about you. More than 30% of credit reports have errors. A recent check on mine revealed addresses I have never lived at, wrong birthday, job experience and some of my son's credit.

Health insurance payments do not show up on credit reports. The best way to find out about your credit is request a free copy from one of the 3 major credit bureaus. Here are the bureaus:

Equifax Credit Information Services
Box 105518
Atlanta, GA 30348
Phone: 877-463-5505 (for Credit Report orders) Phone (800) 685-5000
(for disputes)

Experian
Box 949
Allen, TX 75013-0949
Phone: 888-397-3742 (for Credit Report orders)
Phone (800) 583-4080 (for disputes)

Trans Union Corporation (TUC)
P.O. Box 1000, Chester PA 19022
Phone: (800) 888-4213 (for Credit Report orders)
Phone (800) 916-8800 (for disputes)

If there is anything on your report that is not right, you can dispute it. The disputed company has 30 days to respond. If they don't, they will take it off. Many don't respond. If you still have a problem with an item, you can add a 100 word statement explaining why to your credit report.

Military Discharge

Many candidates who received a less-than-honorable discharge from the military have been able to have their DD214 changed to honorable with the help of Veterans Affairs and advocate groups. Not the Veterans Administration. You have to be 6 months to one year free from any other violations.

Background Investigators

The following information was posted on a firefighter bulletin board. The author is unknown. The information is not verified. Use it only for food for thought:

To those curious to know if we background investigators share information the answer is yes. It does depend on the individual agency however. Some have strict guidelines as to what information can be shared with other investigators. Our purpose as background investigators is to gather information on a potential firefighter, "good" and "negative" information.

A good aggressive background investigator will dig deep into a candidates life and look for things to disqualify a candidate. We as background investigators want to eliminate the bad seeds before they get into a fire department.

This does not mean that a firefighter candidate has to lead the "Perfect Life", just don't do anything stupid to get yourself in trouble.

Remember to use your best judgment when completing a background packet. If you do not put that you were back grounded for another agency and the investigator finds out, you will definitely be disqualified.

To my knowledge, there is no black list that would reveal if you had a background with another department and failed. However, most credit reports show who has made an inquiry on your report. These reports aren't usually obtained by the city though, but by another company on the cities request.

How Honest Are You?

Captain Smith, after going through your program and a coaching session with your son, I am currently in backgrounds with 2 departments in Southern California and on the lists of 4 others. The background investigators for those two departments told me that they would both have psychological evaluations and one will have a polygraph. I want to make sure that I am prepared for these last phases prior to hiring. Your son suggested I call you for advice. Thanks, Randy

After you have jumped through all the flaming hoops you don't want to be caught flat-footed for the remaining steps in the hiring process. It's 3rd down and 2 yards to go for the badge. You want to convert. You want to convert every step of this process the first time through the line, or you could be thrown for a loss, thrown in the penalty box, out of the game, and trying to fight your way back in.

You can spin this anyway you want. But ask yourself if you would you show up without preparing for the written? Not in shape for the physical agility? Have you discovered you just can't wing the oral? Then, why doesn't it make sense to prepare for the remaining portions of the hiring process, the background, psych, poly and medical?

Don't be so naive to believe by the 4 inches between your ears you have an explanation that everything in your past will be overlooked, especially if it's something you weren't required to reveal in the first place. If you do, you might still believe in the Easter Bunny, Tooth Fairy and still leave out cookies and milk for Santa. Come on in said the spider to the fly. Don't take the bait! It's not the department but the background investigators and the psychologist that could take you out. These people are not your friends. They are experts being paid to eliminate you from the process. The deck is stacked against you before you show up.

I get the calls when the background has not gone right for too many candidates. The first words out of their mouth when I pick up the phone is usually, "What do I do now?" I ask them two questions. First, were you honest to a fault leaving no rock unturned? Did you volunteer information that you were not required to give? They usually answer yes to both. That's probably why you failed. The defense rests.

A candidate just called and said the background investigator told him a poly would be given to verify his information. My first question, "Was it listed on the job announcement that there was going to be a poly? No. If it was not included in the job announcement and or they are going to give a poly to everyone else, that's BS.

It's not uncommon for a background investigator or psychologist to say, "Will you submit to a poly to verify your answers? Or, a poly could or will be given at the end of the process." Are they lying? Yep. Wait a minute; I thought everyone was supposed to tell the truth here? I'm not aware of any test where the candidates were held hostage with the threat of a poly being given, when it was not included with the job announcement, and they had to take one.

I know of candidates who were turned down and wanted to take a poly to prove they were telling the truth and they couldn't get one because they would have to give it to everyone else. They often say, I didn't think what I told them was any big deal, but some of those little things that I really didn't have to talk about amounted to causing me big problems in the process. As one candidate said, "Hey, I'm not a bad guy. But I volunteered a little something here and there. By the time they got done with me, they made me look like Charles Manson! "

Those who are critical about what we are saying here probably have never gone through our program and usually don't have a clue what we do. I want candidates to be prepared for each step of the hiring process, where the land mines are and understand the ramifications of the information they present in the process.

You have spent all this time gaining education, experience and training to get this job. You finally get a shot at the badge. You get a conditional job offer. You're ecstatic. You call family and friends. You meet with the background investigator. You think he's your pal. You go for your psych. No big deal right? Then a letter arrives from the department withdrawing their job offer. You're stunned! There has to be a mistake. You want to talk to someone. You had the explanation you knew they would accept. No one will talk to you. You're out of the process. The reason? You walked in flat-footed not prepared for the remaining segments of the hiring process.

As one candidate wrote: As for backgrounds they tell you to be honest. But sometimes being honest can bite you in the ass. When a background is being conducted the only obvious things they could find out are things like your driving, criminal, employment, credit history and people you know. Don't be stupid and write down references that hate you. I've know some good people that should be fireman/cops but get disqualified for being too honest.

You're a free agent. Make sure you prepare for the hiring process in a way that will best put you in a position for a badge.

Clouded Past

Too often I receive calls from candidates after they have spend time, money, education and put their lives on hold for years only to be taken out at one of the hiring stations.

For some they are clouded and doomed before they ever take their first step because of something in their past, job history, criminal record, credit, driving record or domestic violence. They still believe though that they have a shot. They have convinced themselves by the four inches between their ears that they have an explanation about what has happened in the past that will convince those in the hiring process they're the right candidate. So, they pack on all kinds of credentials, degrees, experience, academy, paramedic, etc. thinking this will erase the problems of the past. Then, they don't face it until it's staring them in the face when they start testing or at the next step in the hiring process.

Candidates are often not given the information they really need. They're serious land mines ahead of every candidate. Some more than others. I try to shoot straight saving them time, money, treasure and wasted opportunities. I'm truly saddened when I tell candidate with a clouded past when their chances are of getting hired are not good.

What do you think most say? Well, I'm going on anyway. How many do you think get back to me that they were hired? None!

Those in the hiring process have heard all the stories and look for patterns. The application and background packet will ask you in several different ways, "Have you ever"… Computers don't forget. If you don't put it down and they find it you're immediately eliminated from the hiring process.

Where this can really play out is in the psych evaluation where up to 40% of candidates fail. With problems in your past the doc will ask you some difficult questions that could take the wheels off your wagon.

Too many candidates walk in flat-footed once they're given a conditional job offer and are eliminated in the psych, poly and or medical. These are unchartered waters where you need to be prepared in advance before you show up. These are experts who are being paid to take you out.

Expungement

Many states provide expungement of offenses in varying degrees. What can happen as you get into the hiring process is the background packet will ask in several ways "have you ever", NOTE: Even if you have had an expungement, you still must include the requested information in the background packet.

If you don't list it and they find out, you could be eliminated for not being truthful. Can they find it? I often hear from candidates who have either gone through a diversion programs and or an expungement and were told they didn't have to list the offense on an application because the record had been erased.

From John: Not always true. Many have believed or have been told, "If a potential employer runs a criminal check with DOJ, nothing will appear". Contre my friend, this is not true. If a potential employer runs a criminal check with the DOJ, the conviction will appear, annotated with the phrase (similar) "Case dismissed pursuant to 1203.4."

I've had candidates who have not listed their expungement or diversion programs and it was found causing huge problems. A candidate's ambulance company found it. His first department background didn't, but the agency he really wanted to work for dropped him because they did find it after he did not include it as requested.

Run Like Hell!

We've all heard the stories of people looking for jobs in the fire service, and something they considered a minor problem in their past has kept them out. Let's make sure not to add to that list.

Nothing will get you kicked out of the background check faster than a D.U.I or a domestic violence charge. Both usually involve drugs or alcohol. Don't be stupid. If you are going to be drinking have a designated driver, get a cab, stay in a motel, hell sleep under a bridge, but don't risk throwing all of the time, education, and dreams you had for so long away over a lapse in judgment.

If you've been in a bad relationship, and you don't get along with that person, make it easy on everyone and just stay the hell away from him or her. It is so easy for things to get out of hand and then the police arrive. Later you have to explain it to a background investigator.

"But I'm the best firefighter in the world. If they could just see how good I am they'd take me for sure". I'll tell you right now it ain't so. Don't bet on it. If a department has 100 people to pick from, they don't need to take a chance on someone that has blown it in the past. So don't blow it.

If by chance you end up in one of these situations, you need to do everything you can to minimize the damage. I'm not talking about doing a "Bill Clinton", but anything you have to do to make it right. I know one guy who told his girlfriend he wanted to break up, and she thought they were going to get married. In the scuffle that ensued she had a scratch on her face when the police arrived. Someone was going to jail and it wasn't the bride-to-be. He was being charged with domestic violence, and was in the hiring process for a department. He had to make up, get her to say it wasn't his fault she was scratched, and then keep her happy so she doesn't change her mind.

With all of the preparation we all must go through to get this job, keep in mind to protect your record. Do every thing you can to not have to be explaining things to a background investigator. Prepare before you go out to have a good time, make all of the necessary arrangements before you've been drinking, while your head is clear. If you get into any sticky situations that look like it might go bad, there is one solution that's never failed me RUN LIKE HELL, AND DON'T LOOK BACK.

Good Luck, Captain Rob at mailto:nrtc@sonic.net   http://www.myfireinterview.com

The Call Doesn't Come

You have done everything you can do to get hired. You're in background. Things seem to be moving along. Then nothing is happening. Your calls are not returned. You hear from other candidates below you on the list that they have completed their psych and medical. The academy date has been set. You still can't find anything out.

The letter arrives informing you that you are out of the process. You're devastated. You don't know why. You want to know. Maybe there's something you can do, someone you can talk to convince them you're the guy they need; there has to be a mistake. Just give me a chance to prove myself. Your calls go unanswered.

What do you do now? You feel hurt, angry, frustrated and betrayed. Well, there is not much you can do. If you push too hard you might hurt your chances of getting on somewhere else. Even if you are able to get someone to talk to you, you probably will never really find out why. It just happens sometimes for whatever reason.

My advice. Go through the denial, anger, and depression. Lick your wounds and accept what happened. Then, regroup and get back out there and test again. If you made it this far, chances are you can do it again. Maybe with a better department than you hoped for.

Fire Academy

The purpose of this chapter is to keep you from repeating the errors others have made keeping them from gaining a badge.

Just because you passed the physical agility doesn't mean you are ready for the fire academy. Whether you agree or not, the physical agility has been watered down to be politically correct. Departments know this. So the training division is going to put you through the wringer to make sure you can do the job before you go on line.

Showing up at the academy is not the time to start getting ready. You need to be in shape and hit the ground running. I often get calls from candidates asking what do I do now? They have been let go from the academy. It's tough enough getting a job. Keeping it can be a challenge. If you are let go by one department, it's going to be difficult if not impossible to get another department to take a chance on you.

"The worst mistake is to have the best ladder at the wrong wall." Donald Rumsfeld Secretary of Defense, USA

It's not just the physical part. You have to pass every segment of the academy including the final test to demonstrate you can function in the field. It's not uncommon to have a group of candidates let go in the final two weeks of the academy because they can't master ladder throws, repel or operate the equipment. More than one candidate has been let go because they couldn't start the chain saw, operate the jaws or struggled on the drill ground in the final test.

Nothing will piss of the training staff more than you telling them a better way to do something. How you did it in your FF1 academy, reserve or other department. The only task you need to focus on is how they do it in this department. Training divisions are their own kingdoms. This is not a democracy! You have no time or opinion.

It is devastating to be let go, especially if you have already been through a college fire academy. You have been dropped as your classmates are getting dressed up in their class A uniforms (about the only time they will ever wear it, except for funerals) heading for their badge ceremony.

It starts with instructors from the academy taking you aside and pointing out the problems you are having. If you don't improve, they will meet you again with other members of the training staff and document the meeting. The writing is on the wall if things don't improve. Candidates that get to this point start to panic. This can affect their other skills. Things they already know and have mastered become difficult. Instead of dropping back and taking a different mindset, they start to panic and withdraw. Too many candidates in this situation would rather go below and fall on their sword before they will ask for help.

This is the time to ask for help, extra training, and check in with those who have gone before them. I usually get the call after they have taken the option to resign instead of being fired. My first question is why didn't you call me earlier? Well, I didn't think it was that bad.

Paul told me that six months earlier he and his wife had what I call "the talk." The talk can come whenever situations build up and aren't resolved. It usually blindsides one of the partners. Paul's wife told him he wasn't part of her life. He was always gone with the guys, coaching and/or competing in sports, besides working his firefighter shifts. She didn't think he loved her anymore, and she was losing her love for him, too.

Like most guys, firefighters get many of their needs met at work. We have certainty, uncertainty, variety, significance, discipline, comfort, connection, growth and contribution. Because we go home with our tank pretty full, we aren't always motivated to provide the same vital necessities for our partners.

Paul attempted to make changes. But he made the big mistake of not asking his wife what changes she wanted. And if the changes aren't what the other person needs, it doesn't make any difference what a person does.

So now it was six months later and she still felt the same way. She threatened to leave with their four-year-old son. She stated that because of what had already happened, counseling was not an option.

From our conversation, I suspected that Paul's wife, Tara, was at Dr. Gottman's level four, stonewalling. Had she crossed the line?

I sent Paul home with four chapters from this book. As Tara read the chapter on the five-to-one ratio, she said, "I'm right here," pointing to stonewalling. Had she crossed the line? Reading the other chapters, she said, "Did he write this about you? Because this is exactly what you've been doing."

They went for a walk and discussed the chapters. They talked several times during the next three days. Then Tara caught some hope. Maybe with these new tools, they could work it out. Paul took the time to ask Tara the changes she really needed. He opened a Love Bank Account and started making deposits. He started his transition toward the five-to-one ratio. Three weeks later she felt it was real; Paul wasn't just putting out spot fires, as before. Tara said, "I think I still love you." Three months later they went away to celebrate their anniversary. I saw Paul a month later and he said, "It's never been better."

One year later he told me, "It's still hard work, but well worth it. It's never like before because we have the Nuggets of Life needed to work it out."

This firefighter became a hero in his own home by saving his marriage. Imagine what could happen in your life?

Nugget: Open a Love Bank Account. Use it as an accounting system to transition to the five-to-one ratio.

How: Make the kinds of deposits your partner needs most.

Testimonial from Paul

Two years ago was a terrible year for my family; or so I thought. Actually, my five-year marriage was not as great as I assumed. It all came down on top of me that summer.

Two things happened in our marriage relationship that made me realize something was wrong. First, my wife spent more time and seemed to have more fun with her friends than with me. And second, we hardly ever communicated or did things together. When we did, it was not fun. We had grown apart.

My wife told me that she no longer loved me and I was in danger of losing her to someone else! If I did not become a better husband, father, listener and friend, she was planning on moving out. I was shocked, amazed, hurt and disappointed, but she was right in every way. I did not know who to turn to. I was embarrassed to tell anyone about my problems.

One day while working with Captain Bob, we had a discussion about women. My problems with communicating and listening to my wife surfaced. With Bob's words of wisdom, and passages and excerpts from his book, my wife and I were able to identify the problems in all facets of our relationship. We have become best friends again, and now I can honestly say that we are on the road to salvaging our marriage. It is a continuing process! We must work at it each day.

I realized one thing through this whole ordeal. We both need each other a lot, and if we know how to listen and can communicate better with each other, the better our relationship will be in the long run.

"Being in love means speaking the same language. Loving means speaking the other person's language." – The Rt. Rev. John R. Wyatt

One of my goals is to keep guys out of the penalty box. Oh, we're still going in. We don't even know sometimes what we have done wrong. I try to provide some nuggets of life so guys can stay out the longest time and when they go in it will be a shorter stay.

A guy came out of the phone booth at the firehouse one night. He looked puzzled. I asked Phil what was wrong. He said he just spent an hour apologizing and he still didn't know what he had done.

Understand that firefighters have a higher divorce rate than the average person. Marriage is the only war where you sleep with the enemy. A San Francisco command class instructor recently said the divorce rate for SFFD is 75%. He said, "If you want to get married and be a firefighter find someone you hate and buy them a house."

"Love is grand. Divorce starts at a 100 grand."

Nothing is worst than being up half the night on calls on get away day and coming home for your days off and the bride is already mad.

It's no fun in the penalty box. There is no romance. Not even maintenance sex. This might help:

Five Secrets for Successful Relationships

Simple Tools to uncomplicate lives

Recognize, understand, and accept that we are dramatically, dramatically different in incalculable ways. We are constantly judging each other by our own quite different standards. What may seem crystal clear to you probably is not to your partner.

Find out what makes that person in your life feel special or loved. Open a love bank account and start making deposits of those things that make that person feel loved. You will receive the interest and dividends from the account.

Transition from criticism, defensiveness, contempt, and stonewalling to a ratio of five positive moments to each negative moment in your relationship. The love bank can be the accounting system for the five-to-one ratio. The one negative moment is just as important as the five. The short-term misery will clear the air and add newness to the relationship.

Know what you want. Condense it down to 15 seconds or less. Is this realistic for the person you are with? If not, go back to the beginning; present what you want to your mate. Have your partner repeat back what they heard. You might have to go back and forth a few times before it gets translated. Make adjustments where needed. Write the expectation down so there is no amnesia later.

Plan evenings out and trips. Then, follow through and do them. If you don't plan, you won't go. It's not optional. For your own mental health, it's mandatory. Schedule them on a calendar. The anticipation before and the memories after are priceless. Go first class once in a while. If you don't, your heirs will. Adopt the philosophy that if you are not on a trip, you are planning the next one.

"Most people don't change because they see the light, they usually feel the heat."

The grass is greener on the other side of the fence. It's just as hard to chew. You've got to mow it, too. It's just different grass.

This is a response from Mary:

Fire Captain Bob, Thank you so much for your time, I was acting a bit selfish and will try to be more understanding and supportive. I didn't know that this City was the toughest academy in the nation, but it sure did seem like it! I guess I was right! Because you had to hear all about my sad sob story, I should update you: my husband is doing better and his spirits have been uplifted since he received a 95 on his midterm! I will take all your advice and put it to work now, where can I find your books. I would like to read them thank you so much! The world needs more people like you. — Mary

Thank you for your gracious words. I'm glad you took the risk to contact me. Glad to help. You can find the books at eatstress.com

New Rookie

I talked to a devastated candidate at a written test. This paramedic had been hired with four other medics by a good fire department. After four months he was fired. He said he thought things were going fine. Then, the captain started telling him that the other firefighters didn't like some things he was saying, started counseling and documenting him for not taking down the flag, rolling up the hose, etc. He said he was busy doing other assignments. The writing was on the wall.

I asked him what the other new rookies were doing? He said they were too busy kissing ass. My only reply was, "I hope you learned that if you were too busy kissing ass, you wouldn't be trying to get another job!"

What you do when you first start out will set your reputation and follow you throughout your career. If you don't start out on the right foot, they will show you the door. The crew already knows more about you before you show up than you think.

Use these standards during station visits, your interview process, and as a new rookie to demonstrate you already know what to do when hired:

You're a snott nose rookie. Keep your mouth shut. Be cordial, friendly and humble. You have no time or opinion until you earn it. You can't force it. That will come with a lot of calls and a few fires.

Cell phones are causing problems for candidates and rookies. I can't believe the stories I'm hearing. Candidates are carrying their cell phones to written tests. A candidate was in a department academy and his cell phone starts to ring. He told the training officer, can you hold on a minute, I have a call. Yeah, right. The training officer told the class the next time he hears a cell phone go off, they were going to play who can throw the cell phone the furthest.

On an emergency call, the BC was trying to raise dispatch without success on the radio. The rookie took his cell phone, speed dialed dispatch and handed his cell phone to the BC. Cute? Smart? Innovative? That's not the reception he received.

Rookies are carrying their cell phones on duty. Their phone rings, they answer it and go right into cell yell with their friends and relatives. Wives, girl friends and dysfunctional others call all day long with important stuff and to do pillow talk. Cell phones are ringing in locker rooms. Some try to be cool by putting their cell phones on vibrate or stun. Even though they might not answer them when they go off, they still pick them up to check the caller ID or the text message. Then when they think no one is looking, they slip off and return the call or are constantly texting. THIS IS DUMB! These are not part of your emergency issue.

This will not get you off on the right foot. Big clue here. Leave the electronic leashes off and in your vehicle, along with your piercings, until a time where all your duties are complete. No matter what you might think and how friendly everyone seems to be, you are being watched! It could hurt you big time.

If you have an emergency situation, ask your officer if you can carry your phone because you are expecting an emergency call.

Call your new captain before your first shift and ask if he wants you to bring anything in. Bring a peace offering of donuts, gourmet coffee or desert your first day. Homemade is best. Arrive early and ask the off-going firefighter what you should know at that station. Your new captain should meet with you to outline his expectations. If not, ask him.

Unless you're told differently, put up and don't forget to take down the flag. If the phone or the doorbell rings make sure you're the first one running to answer it. There will be certain duties on each day of the week. Tuesday could be laundry day, Saturday yards. Keep track. Stay busy around the station. Always be in a clean proper uniform. Always be ready to get on the rig and respond.

Check out the gear on the rig each morning. Make sure the O$_2$ gauge and the reserve bottle shows enough to handle a long EMS call.

Firefighters usually have "Their" place to sit at the table and in front of the TV. Don't hog the newspaper. The off-going shift has the first crack at the newspaper. You probably have probation tests. Don't park yourself in front of the TV; you have a test coming up. Stay busy. No matter what the atmosphere, you're being watched.

Although you might be a good cook, don't volunteer to cook until asked or rotated in. Make sure your meals are on time. The old adage "Keep them waiting long enough and they will eat anything" doesn't apply here. Be the last one to serve your plate. Don't load up your plate the first time around. Wait to go for seconds.

Always have your hands in the sink doing the dishes after a meal. Be moving out with the garbage and mopping the kitchen floor after each meal.

Learn how to help the officer complete response reports.

Don't tell jokes until you're accepted.

Don't gossip.

Don't play "Your" music on the radio. Don't be a stupid generation X'er or Y'er and always ask why when told to do something. Help others' with their assignments when you finish yours.

Ask how you're doing. Volunteer for assignments. Keep track of these to present at your evaluations.

Don't start pulling hose and other equipment at a scene until the captain tells you.

Always get off the rig before it backs up. Stand to the rear side to guide the rig. Never turn your back on the backing up rig.

It's not uncommon to move to one or more stations during your probation. At your new station, don't act like you already have time. Unfortunately, you have to start all over again as the new rookie.

You will have an elated feeling rolling out on your first calls. There is nothing like it. It could last your whole career. Enjoy and savor it. You earned it. You're the last of America's Heroes.

I miss it.

Hope is the Anchor for the Soul!

You can't figure out why you got a 100% at LA City and didn't get the letter to move on in the process. Didn't score high enough on the written in Phoenix to get an oral. Waiting for the letter for band II for LA County. Trying to figure out if it's worth it to pony up another $175.00 and drive all the way to take the CPAT physical agility. You believed the psychologist in Seattle when he told you the psych interview wasn't designed to eliminate candidates to get you to open up; then you were out of the process afterward. You don't know what to do next.

Many candidates get bummed our and lose hope in the process. "Hope is the anchor for the soul."

Understand getting this job is a process. You need to take it one step at a time. If you aren't going further in the process, you have to start asking yourself why. Try to identify what part of the process you're getting stuck in. Is it the written, physical agility, oral, medical, psychological, background? You need to be prepared before you take the next step in this process. Whatever area it might be, take that area and slice it apart, gain the necessary resources to be able to make it to the next step. If you don't nothing will change. You need to hang in there just a little longer. Stay motivated when it seems the system is not working.

You have to be the energizer bunny and keep going, and going, and going if you want to get the job of your dreams. You have to fly, drive, beg, borrow and grovel to make it happen.

Keep a vision of seeing yourself in that job. My son Rob did this when he was struggling to follow in his father's footsteps to become a firefighter. His vision was, "Seeing someone pin my badge on me?" He got that job. And on that magic day, emotionally, I had the honor of pinning that badge on my son. I felt he had received a degree and career in one swoop. Because, he won the lottery and he didn't even buy a ticket!

We have a small flower garden in front of our house. It's my piece of dirt. It's a color spot I enjoy. It didn't just happen. It took work, experience, and tools. First, I had to prepare the ground, then wait for the frost to end, select the variety of plants from what was available, and, finally, use Vitamin B while transplanting to insure that the new plants would take hold.

If there were any existing blooms on the plants when I transplanted them, I had to snap them off because they would sap so much energy and nourishment they would threaten the chances of the plant making it through the shock period. Once the plant had recovered and was producing new flowers I had to continue to remove them as they matured or else they would have gone to seed and the plant would have stopped producing new flowers. This process also strengthened and filled out the plant.

There's a very similar experience in nourishing our desires to be a firefighter. If we are not constantly cultivating, fertilizing, and watering our dream, it will become seedy, dry and withered.

At one point when I first started my garden the plants weren't producing many flowers. I checked everything. I had fertilized, watered and had removed the mature flowers. All the necessary steps had been followed. I asked my wife Harriet what was wrong. The answer: I needed to water more and cultivate the ground to allow the water to penetrate the soil to get to the roots. Within days after more watering it was like the horn of plenty. Color was everywhere. Again, like our dream to be a firefighter, we might have the necessary elements but in the wrong amounts for the harvest to take place.

Many people desire the nurturing relationships that produces a garden with a shining badge. They want their garden to attract the elusive butterfly of peace and happiness that can get them the badge. But just like trying to catch a beautiful butterfly that seems to dart and flutter away, it escapes their grasp right at the last minute because they probably don't have the necessary "Nuggets" of life to catch it. It was that way for me. It was the "Nuggets" of life that I acquired on this journey and my experience in using them and putting things in place that produced the job of my dreams.

With all the frustration, anxiety, depression, and retooling that went on, I thought things would never really change. I can't tell you exactly when it happened, but all the pieces of the puzzle came together. I wasn't chasing the butterfly. I was just standing calmly and this beautiful butterfly landed on my shoulder. It turned into the badge that my wife pinned on my chest. I felt like the turkey when the red bulb pops out. I was done, and I knew it.

This is my prayer for you, friend.

"Hope is hearing the music of the future . . . Faith is dancing to it today!"

Please let us know when you receive your badge. If you have thoughts, comments or ideas about Becoming a Firefighter, I would enjoy hearing from you.

If you would like me to speak to your group or organization, please contact to me at:

Fire "Captain Bob"
Phone: (888) 238-3959
E-mail: Mailto: Robert.smith19@comcast.net or mailto:captbob@eatstress.com
Web site: http://www.eatstress.com

Captain Rob's web site: http://www.myfireinterview.com e-mail: mailto:captrob@sonic.net

You can learn more about our Entry Level, Promotional Programs and other products from our web site: http://www.eatstress.com or by calling toll free (888) 238-3959.

You can stay ahead of the curve by signing up for our FREE FireZine e-mail newsletter off our web site here: http://eatstress.com/firezine_signup.htm

Remember, "Nothing counts 'til you have the badge . . . Nothing!